Diss. ETH Nr. 16817

Growth of CsCl-type domains on icosahedral quasicrystal Al-Pd-Mn

A dissertation submitted to the
ETH Zurich

for the degree of
Doctor of Sciences ETH Zürich

presented by

YVES WEISSKOPF
Dipl. Phys. ETH
born on February 2, 1978
citizen of Pratteln, BL

accepted on the recommendation of

Prof. Dr. M. Erbudak, examiner
Prof. Dr. D. Pescia, co-examiner
Prof. Dr. L. Schlapbach, co-examiner

2006

Bibliografische Information der Deutschen Nationalbibliothek

Die Deutsche Nationalbibliothek verzeichnet diese Publikation in der
Deutschen Nationalbibliografie; detaillierte bibliografische Daten sind
im Internet über http://dnb.d-nb.de abrufbar.

ISBN 3-8325-1420-1

Logos Verlag Berlin
Comeniushof, Gubener Str. 47,
10243 Berlin
Tel.: +49 030 42 85 10 90
Fax: +49 030 42 85 10 92
INTERNET: http://www.logos-verlag.de

We have no choice left but betwixt
a false reason and none at all. [1]

Es bleibt uns nur die Wahl zwischen
falscher Erkenntnis oder gar keiner. [2]

David Hume, 1711 − 1776

Contents

Preface

This dissertation is based on research done in the group of Prof. Dr. M. Erbudak at the Laboratory for Solid State Physics, Department of Physics, Swiss Federal Institute of Technology Zurich (Eidgenössische Technische Hochschule Zürich). Parts of it, in particular Figs. 3.1 − 3.5, 4.1 − 4.4, 5.1, 5.3, and 5.5, are reprinted from Surface Science 578, Y. Weisskopf, R. Lüscher, M. Erbudak, Structural modifications upon deposition of Fe on the icosahedral quasicrystal Al-Pd-Mn, pp. 35 − 42, Copyright (2005); Surface Science 600, Y. Weisskopf, M. Erbudak, J.-N. Longchamp, T. Michlmayr, Ni deposition on the pentagonal surface of an icosahedral Al-Pd-Mn quasicrystal, pp. 2594 − 2599, Copyright (2006); and Surface Science, Y. Weisskopf, S. Burkardt, M. Erbudak, J.-N. Longchamp, The quasicrystal-crystal interface between icosahedral Al-Pd-Mn and deposited Co: evidence for the affinity of the quasicrystal structure to the CsCl structure, in press; with permission from Elsevier.

The structure of this dissertation is as follows: Chapter 1 presents an introduction to topics subsequently discussed. Experimental details are provided in Chapter 2. Experimental results of the growth of Co, Ni, and Fe deposited onto the pentagonal surface of icosahedral Al-Pd-Mn are presented and discussed in the Chapters 3, 4, and 5, respectively. The Chapters 4 and 5 refer to results presented in Chapter 3. Chapter 6 provides a comparison of the different growth modes, a conclusion of the results, and a short outlook.

Many people have contributed, directly or indirectly, to the success of this dissertation and I am much obliged to all of them. In particular, I would like to express my gratitude to my parents Christine and Thomas Weisskopf, my sisters Fabienne and Eliane Weisskopf, as well as Heidi Windlin for their enormous support; to my doctoral adviser Prof. Dr. Mehmet Erbudak for his generous assistance and the nice time we had working together; to Prof. Dr. Danilo Pescia for his support and his interest in this work; to my co-workers Dr. Thomas Flückiger, Dr. Rouven Lüscher, Jean-Nicolas Longchamp, Sven Burkardt, Thomas Michlmayr, Dr. Andreas Vaterlaus, Dr. Urs Ramsperger, and Niculin Saratz for their miscellaneous assistance and the pleasure working together; to Prof. Dr. Muhittin Mungan, Paolo Moras, and Dr. Michael Hochstrasser for the good collaboration; to Dr. Esther Belin-Ferré for providing me with the quasicrystal sample; to Dr. A. Refik Kortan for his support; to Prof. Dr. Louis Schlapbach for his interest in this work; and to Sofia Deloudi, Prof. Dr. Walter Steurer, and Dr. Thomas Weber for enlightening discussions.

Zusammenfassung

In dieser Dissertation wird die Bildung der Übergänge zwischen der fünffach-symmetrischen Oberfläche eines Al-Pd-Mn Quasikristalls, welcher eine nominelle Zusammensetzung von $Al_{70}Pd_{20}Mn_{10}$ aufweist, und aufgedampftem Co, Ni sowie Fe behandelt. Die chemische Zusammensetzung in einem oberflächennahen Bereich wurde mittels Auger-Elektronenspektroskopie ermittelt. Die Oberflächenstruktur wurde mit Hilfe niederenergetischer Elektronendiffraktion und der Abbildung von Sekundärelektronen untersucht. Longitudinale magneto-optische Kerr-Effekt-Messungen wurden durchgeführt um magnetische Ordnung in der Oberflächenebene zu detektieren. Rasterelektronenmikroskopie mit Polarisationsanalyse der Sekundärelektronen wurde zwecks Nachweis magnetischer Ordnung in der bzw. senkrecht zu der Oberflächenebene durchgeführt.

Für eine Submonolage von abgelagertem Co bildet sich eine atomare Schicht an der Oberfläche, welche aus fünf AlCo-Domänen mit Abmessungen im Nanometerbereich besteht. Die Domänen besitzen eine Struktur, welche einer CsCl-artigen (110)-Oberflächenstruktur entspricht, und sind um 72° zueinander gedreht. Ein Vermischen von Co mit dem Substrat senkrecht zur Oberfläche wird nicht beobachtet. Die Orientierung der AlCo-Domänen bezüglich dem Substrat ist durch die optimale Anpassung der gemittelten Strukturen gegeben, welche jedoch vier unterschiedliche Domänenausrichtungen erlauben würde. Die beobachtete Ausrichtung ist auf Grund von ausrichtungsabhängigen Oberflächen- und Domänen-Domänen-Grenzflächenenergien bevorzugt. Co in der kubisch raumzentrierten Struktur wächst epitaktisch auf den AlCo-Domänen für weitere Co-Ablagerungen bis zu einer Schichtdicke von etwa 3 Monolagen. Für grössere Ablagerungen bildet sich auf den Co-Domänen eine modifizierte Oberflächenstruktur bestehend aus eindimensionalen Atomreihen. Für Ablagerungen von mehr als etwa 1,5 Monolagen zeigt die gesamte Co-Schicht magnetische Ordnung in der Oberflächenebene.

Ni vermischt sich mit der Substratoberfläche in der ersten Phase des Wachstums und es bildet sich eine ungeordnete Al-Mn-Ni-Schicht. Für weitere Ni-Ablagerungen wandert Al vom Substrat zu fünf Al-Ni-Domänen, welche auf dieser Schicht wachsen. Diese Domänen besitzen eine Oberflächenstruktur, welche einer CsCl-artigen (110)-Oberfläche entspricht, sind um 72° zueinander gedreht und haben Abmessungen im Nanometerbereich. Ähnlich wie bei Co resultiert die Orientierung der Domänen bezüglich dem Substrat aus der optimalen Anpassung der gemittelten Strukturen. Dies weist darauf hin, dass die Al-Mn-Ni-Schicht, welche Strukturinformation übermittelt, die selbe Struktur und Orientierung wie die Al-Ni-Domänen annimmt. Für weitere Ablagerungen wächst auf diesen Domänen Ni, welches möglicherweise etwas Al enthält, in der kubisch raumzentrierten Struktur. Für Ablagerungen von mehr als etwa 4 Monolagenäquivalent bildet sich eine Reihenstruktur auf den Domänenoberflächen. Ni-Schichten weisen magnetische Ordnung senkrecht zur Oberflächenebene auf.

Ähnlich wie Ni vermischt sich Fe mit dem Substrat in der ersten Phase des Wachstums und es bildet sich eine ungeordnete Al-Fe-Mn-Pd-Schicht. Für weitere Fe-Ablagerungen wandert Al vom Substrat zu einer Al-Fe-Schicht, welche auf der Al-Fe-Mn-Pd-Schicht wächst und für genügend hohe Fe-Bedeckungen von etwa 4 Monolagenäquivalent aus fünf CsCl-artigen Domänen besteht. Diese zeigen ihre (110)-Oberflächen, sind um 72° zueinander gedreht und haben Abmessungen im Nanometerbereich. Ähnlich wie in den Fällen von Co und Ni ist die Orientierung der Domänen bezüglich dem Substrat durch die optimale Anpassung der gemittelten Strukturen gegeben. Die Al-Fe-Mn-Pd-Schicht scheint die selbe Struktur und Orientierung wie die Al-Fe-Domänen anzunehmen. Für weitere Ablagerungen wächst auf diesen Domänen Fe in der kubisch raumzentrierten Struktur, wobei das Wachstum dreidimensionaler Natur ist und diese Schicht möglicherweise etwas Al enthält. Für Ablagerungen von mehr als etwa 8 Monolagenäquivalent weisen die Fe-Domänen eine geringe Schieflage auf. Magnetische Ordnung in der Oberflächenebene tritt für Fe-Bedeckungen von mehr als etwa 4 Monolagenäquivalent auf. Eine nichtmagnetische Schicht von etwa 2,5 Monolagenäquivalent bildet sich am Übergang zwischen Fe und dem Substrat auf Grund der Vermischung.

Das Entfernen von Co- und Ni-Ablagerungen von weniger als etwa einer bzw. 2 Monolagen oder von Fe-Ablagerungen von weniger als etwa 4 Monolagenäquivalent mittels Sputtern offenbart eine Struktur, welche aus fünf CsCl-artigen AlPd-Domänen besteht, welche ihre (113)-Oberflächen zeigen. Werden grössere Ablagerungen entfernt, sind fünf CsCl-artige AlPd-Domänen zu beobachten, welche ihre (110)-Oberflächen zeigen. In beiden Fällen sind die Domänen um 72° zueinander gedreht und bezüglich bestimmten Symmetrieachsen des Substrats ausgerichtet. Die Bildung von AlPd auf Al-Pd-Mn wird durch Ionenbeschuss hervorgerufen. Die Ausrichtung der Domänen in der Oberflächenebene wahrt die Substratsymmetrie und die grundlegende Orientierung der Domänen ist durch die optimale Anpassung mit der periodischen, diskreten mittleren Struktur des Substrats gegeben, welche vier verschiedene Ausrichtungen von Domänen erlaubt. Eine dieser Ausrichtungen wird nicht beobachtet, während die AlPd-Domänen, welche ihre (113)-Oberflächen zeigen, mit zwei sich ähnlichen Ausrichtungen übereinstimmen können. Der strukturelle CsCl(110)-Charakter des Quasikristall-Kristall Übergangs, welcher sich für genügend dicke Ablagerungen etabliert hat, bevorzugt während des Sputterns die Bildung von AlPd-Domänen, welche ihre (110)-Oberflächen zeigen.

Abstract

In this dissertation, the formation of the interfaces between the fivefold-symmetry surface of an icosahedral Al-Pd-Mn quasicrystal, having a nominal bulk composition of $Al_{70}Pd_{20}Mn_{10}$, and evaporated Co, Ni, as well as Fe is discussed. The chemical composition in a near-surface region has been determined by means of Auger electron spectroscopy. The surface structure has been investigated using low-energy electron diffraction and secondary-electron imaging. Longitudinal magneto-optical Kerr effect measurements have been performed in order to detect in-plane magnetic ordering. Scanning electron microscopy with polarisation analysis of secondary electrons has been carried out for the purpose of detecting in-plane as well as out-of-plane magnetic ordering.

For submonolayer Co deposits, an atomic layer consisting of five AlCo domains with nm dimensions is formed at the surface. The domains possess a structure derived from a CsCl-type (110) surface and are rotated by 72° with respect to each other. No intermixing of Co with the substrate normal to the surface is observed. The orientational relationship between the AlCo domains and the substrate is determined by the optimum matching of the average structures which would, however, allow four different alignments of domains with respect to the surface. The observed alignment is favoured due to alignment-dependent surface and domain-domain interface energies. Co in the body-centred cubic structure epitaxially grows on the AlCo domains for further deposition up to a film thickness of about 3 monolayers. For larger deposits, a modified surface structure is formed on the Co domains which consists of an arrangement of one-dimensional atomic rows. For deposits of more than about 1.5 monolayers, the entire Co film exhibits in-plane magnetic ordering.

Ni intermixes with the substrate surface at the initial stage of growth forming an unordered Al-Mn-Ni layer. For further deposition, Al migrates from the substrate to five Al-Ni domains which grow on this layer, possess a surface structure derived from a CsCl-type (110) surface, are rotated by 72° with respect to each other, and have nm dimensions. Similar to Co, the orientational relationship between the domains and the substrate results from the optimum matching of the average structures. This suggests that the Al-Mn-Ni layer, which transmits structural information, assumes the same structure and orientation like the Al-Ni domains on which Ni in the body-centred cubic structure, potentially containing some Al, is growing for further deposition. For deposits of more than about 4 monolayers, a row structure is formed on the domain surfaces. Ni films exhibit out-of-plane magnetic ordering.

Similar to Ni, Fe intermixes with the substrate at the initial stage of growth forming an unordered Al-Fe-Mn-Pd layer. For further deposition, Al migrates from the substrate to an Al-Fe layer which grows on the Al-Fe-Mn-Pd layer and consists, for sufficiently high Fe coverages of about 4 monolayer equivalents, of five CsCl-type domains which expose their (110) faces parallel to the surface, are rotated by 72° with respect to each other, and possess nm dimensions. Similar to the cases of Co and Ni, the orientation of the domains with respect to the substrate is determined by the optimum matching of the average structures. The Al-Fe-Mn-Pd layer appears to assume the same structure and orientation like the Al-Fe domains on which Fe in the body-centred cubic structure, exhibiting a three-dimensional growth mode and possibly containing some Al, grows for further deposition. The Fe domains exhibit a small tilt for deposits of more than about 8 monolayer equivalents. In-plane magnetic ordering is found for Fe coverages of more than 4 monolayer equivalents. A magnetic dead layer of about 2.5 monolayer equivalents is formed at the interface between Fe and the substrate due to intermixing.

Removing Co and Ni deposits of less than about 1 and 2 monolayers, respectively, or Fe deposits of less than about 4 monolayer equivalents by means of sputtering reveals a structure consisting of five CsCl-type AlPd domains exposing their (113) faces. If larger deposits are removed, five CsCl-type AlPd domains exposing their (110) faces are observed. In both cases, domains are rotated by 72° with respect to each other and aligned with particular symmetry axis of the substrate structure. The formation of AlPd on Al-Pd-Mn is ion-bombardment-induced. The in-plane rotational alignment of domains conserves the substrate symmetry on a global scale and the fundamental orientation of domains is determined by the optimum matching with the periodic, discrete average structure of the substrate allowing four different alignments of domains. One of these alignments is not observed, while the AlPd domains exposing their (113) faces may correspond to two similar alignments. The structural CsCl(110) character of the interface, established for large enough deposits, favours the formation of AlPd domains exposing their (110) faces during sputtering.

Chapter 1

Introduction

Until the discovery of the quasicrystalline state in 1982 [3], solids were generally classified according to their structure as amorphous or crystalline. In the case of amorphous structures, the range over which translational and orientational correlations decay to zero is finite, while a crystal can be generated by periodic translations of a unit cell resulting in a regular arrangement of atoms having long-range orientational[1] and periodic translational order[2].

In the 1780s, René-Just Haüy postulated the law of rational indices according to which each face of a crystal can be unambiguously characterised by three, in general, small integers [5,6]. In 1931, this law was first questioned by Goldschmidt et al. who investigated the mineral calaverite ($Au_{1-p}Ag_pTe_2$, p < 0.15) [7]. They were not able to index the well-defined faces of their samples with three integers. In 1964, Brouns et al. reported that they need to introduce an additional reciprocal dimension in order to index the diffractograms of γ-Na_2CO_3 [8].

Calaverite [5, 9] and γ-Na_2CO_3 [5, 10] are both incommensurately modulated structures, i.e., they have a crystalline structure with a quasiperiodic[2] modulation [11]. Incommensurately modulated structures belong to the category of incommensurate crystals which additionally includes incommensurate composite structures. An example for an incommensurate composite structure is the high-pressure phase of Sc (Sc-II) [12]. It consists of crystalline subsystems with mutually incommensurate structures [11]. Models of incommensurate crystals are based on higher-dimensional crystallography which was developed by de Wolff et al. [13]. Although aperiodic in three dimensions, incommensurate crystals still possess crystallographic orientational symmetry [14].

Quasiperiodic crystals termed quasicrystals are a well-defined ordered solid phase with long-range orientational and quasiperiodic translational order[2] [4]. The same description also holds true for incommensurate crystals. However, in contrast to structures generally referred to as quasicrystalline, quasiperiodicity and orienta-

---

[1]A structure possesses long-range orientational order if the bond angles between neighbouring atoms or clusters have long-range correlations and are oriented, on average, along a set of directions which define the orientational order [4].

[2]A structure possesses periodic (quasiperiodic) translational order if its density function is periodic (quasiperiodic). A function is referred to as quasiperiodic if it is a sum of periodic functions with periods of which at least two are incommensurate with respect to each other, i.e., their ratio is irrational. [4]

tional symmetry are decoupled in the case of incommensurate crystals [4]. Furthermore, quasicrystals can possess arbitrary orientational symmetry. Similar to incommensurate crystals, they can be described as periodic hypercrystals in n-dimensional hyperspaces. A direct space structure results from cutting a hypercrystal with physical space having an irrational slope with respect to the lattice of the hypercrystal [15]. For two-dimensional, decagonal (d-)quasicrystals, which represent a periodic stacking of quasiperiodic layers, n is equal to 5, while for three-dimensional, icosahedral (i-)quasicrystals n is equal to 6. The free energy of aperiodic crystals, i.e., incommensurate crystals as well as quasicrystals, is invariant regarding translations of the physical space, cutting the hypercrystals, perpendicular to itself [16]. Due to this invariance, excitations named phasons exist in quasicrystals giving rise to atomic rearrangements. They cause most of the diffuse scattering observable in coherent X-ray scattering experiments [16, 17]. Although quasiperiodic in physical space, quasicrystals possess a periodic average structure which is discrete, i.e., the projection of the quasicrystal structure into a unit cell of the periodic average structure does not fill it densely [18]. In the case of the three-dimensional Penrose tiling, the average structure has a face-centred cubic unit cell which is decorated with discrete triacontahedra defining the maximum possible distance of an atomic position from a lattice site.

Approximants are crystalline structures closely related to quasicrystals. Their direct space structures are generated from hypercrystals which are related to quasicrystals by a cut with the physical space having a rational slope relative to the lattice of the hypercrystal and approximating the irrational slope used in order to generate the quasicrystalline structures [15]. As a result, the atomic arrangement within their unit cells approximates the local atomic structures of corresponding quasicrystals [19].

Diffraction patterns of quasicrystals consist of a set of Bragg peaks which densely fill reciprocal space, since peaks are found at positions of linear combinations of the incommensurate reciprocal-space periods [4]. However, in experiments only spots with intensities above a certain threshold are observed. The rotational symmetry of patterns reflects the orientational order of a quasicrystal. The set of all diffraction vectors belonging to a quasicrystal forms a $\mathbb{Z}$-module of rank n [15].

Quasicrystals are typically binary, ternary, or quaternary alloys of which many are based on Al [20]. Two prominent examples of stable quasicrystalline alloys are d-Al-Co-Ni and i-Al-Pd-Mn. Like other stable quasicrystal phases, they can be grown using conventional single-crystal growth techniques including the flux-growth, Bridgman, and Czochralski techniques [21]. Quasicrystalline phases usually only exist in a relatively narrow range of an alloy system. In the case of Al-Pd-Mn, the initial composition of the melt has to be in the range of $71 - 78$ at.% Al, $15 - 22$ at.% Pd, and $4 - 10$ at.% Mn in order to obtain a pure i-phase [22].

An alloy structure is stable if the free energy $F = U - TS$ exhibits a minimum, whereas U is the internal energy, T the temperature, and S the entropy [23]. For most solid phases, the free energy is minimal for a minimum of the internal energy. However, in the case of quasicrystals, the contributions of internal energy and entropy to the stability are still under investigation (see, e.g., Ref. [24]).

Already in 1986, Bancel and Heiney proposed that the stability of quasicrystals could be explained by the minimisation of the internal energy [25]. Electronegativity differences, atomic size ratios, and band-structure considerations should be taken into account in particular. However, electronegativity differences between quasicrystal constituents are generally to small in order to contribute to the stability of the quasicrystalline phase. Furthermore, no correlation between atomic sizes and lattice parameters of i-transition-metal alloys is found which suggests that the electronic structure plays the key role in the structure determination. Based on these considerations, Bancel and Heiney argued that the i-phase is stabilised by a Hume-Rothery-like mechanism [26]. Accordingly, the energy of an alloy system is lowered if, at a given average number of conduction electrons per atom, the system assumes a structure which places a Brillouin-zone boundary in contact with the Fermi surface giving rise to a structure-induced minimum in the electronic density of states at the Fermi level [27], a so-called pseudogap. In fact, pseudogaps have been observed for several quasicrystals [28,29] despite of the lack of Brillouin-zones in quasiperiodic structures. However, in correspondence to the construction of Brillouin-zones in the case of crystals, planes perpendicular to reciprocal-lattice vectors related to intense diffraction spots define a quasi-Brillouin-zone. Indeed, Rotenberg et al. were able to describe the quasiperiodic potential related to dispersing states found in the density of states of d-Al-Co-Ni with a few reciprocal-lattice vectors [30]. Although Bloch's theorem is not valid for quasicrystals, extended, so-called critical eigenstates and corresponding continuous spectra exist as a consequence of Conway's theorem, which states that any local pattern of radius d in an ideal quasicrystal can be found in a nearby region within a distance $2d$ [31]. Therefore, the probability that an eigenstate belonging to a cluster tunnels to the next cluster where it assumes an identical form, expect for a damping factor, is finite [32].

Although the above considerations regarding the minimisation of the internal energy agree with many experimental findings, studies regarding the temperature dependence of diffuse scattering suggest that the configurational entropy is decisive for the stabilisation of quasicrystals [16,24]. As a consequence, quasicrystals would be stable only at elevated temperatures and a transformation to a crystalline state at lower temperatures is not observed only due to limited kinetics [33].

Quasicrystals possess extraordinary physical properties, in particular with respect to their constituents, like high electrical resistivity [34], low thermal conductivity [35], high hardness [36], and low surface free energy [36,37]. However, these properties are generally not unique to quasicrystals. They have also been observed in some crystalline alloys, in particular in approximants. A pseudogap, e.g., which has been associated with properties like the low surface free energy [38], is not a specific property of quasicrystals [28].

Three different methods are generally utilised in order to prepare quasicrystalline surfaces in ultrahigh vacuum [39]. In the first method, polished surfaces are cleaned by cycles of sputtering with noble-gas ions and subsequent annealing. Significant changes occur at the surface due to ion-bombardment-induced modifications of the chemical composition[3], segregation, and evaporation. Scanning

---

[3]Regarding ion-bombardment-induced changes in the chemical composition of alloy surfaces see, e.g., Ref. [40].

tunnelling microscopy measurements suggest that the tenfold (10f)-symmetry surface of d-Al-Co-Ni [41] and the fivefold (5f)-symmetry surface of i-Al-Pd-Mn [42] of sputter-annealed samples are bulk-terminated. In the second method, the oxide layer present at a polished surface is thermally evaporated. Again, the surface is modified because of segregation and evaporation. In the third method, quasicrystals are cleavage fractured. The surface cannot equilibrate via diffusion and a rather rough surface results.

Due to the fundamental structural difference between quasicrystals and crystals, quasicrystal-crystal interfaces are of special interest. They have been generated by means of (a) sputtering quasicrystalline surfaces, (b) depositing foreign atoms on quasicrystals, and (c) evaporating quasicrystalline films on crystals.

(a) As a consequence of the ion-bombardment-induced modification of the chemical composition in a near-surface region, the quasicrystal structure is destabilised. In the case of the 10f-symmetry surface of d-Al-Co-Ni, CsCl-type[4] $Al_{50}(CoNi)_{50}$ domains exposing their (110) faces parallel to the surface and aligned with particular symmetry axes of the substrate are established by means of Ar^+-ion sputtering [43,44]. Similarly, CsCl-type AlPd domains are formed on i-Al-Pd-Mn by sputtering the pentagonal surface with Ar^+ ions [45–49]. However, unlike d-Al-Co-Ni, two domain orientations are observed. Both CsCl-type domains exposing their (110) as well as (113) faces parallel to the surface are detected. In addition, both single-domain overlayers [45–47], breaking the rotational symmetry of the substrate, as well as overlayers consisting of five domains rotated by $72°$ ($= 360°/5$) with respect to each other [45, 48], conserving the substrate symmetry on a global scale, are observed in both cases.

(b) Over the past years, a great variety of adsorbates has been deposited on quasicrystalline surfaces (see Table 1.1). Among other phenomena like, e.g., heterogeneous nucleation [37,51,53,57] and pseudomorphic growth for monolayer (ML) coverages [60,80], the growth of crystalline domains with nm dimensions and orientations mediated by the substrate symmetry has been observed for the deposition of foreign atoms [37, 50–56, 58, 65–68, 72, 73, 75, 81]. In the case of Al deposition on the pentagonal surface of i-Al-Pd-Mn [54, 55], e.g., five face-centred cubic Al domains are formed. At a substrate temperature between 250 and 300°C during evaporation resulting domains expose their (111) faces perpendicular to threefold (3f)-symmetry axes of the substrate 37.38° away from the surface normal and have the $[01\tau]$ direction perpendicular to the surface. For a substrate temperature of $200 - 250$°C, growing domains have their [111] axes predominantly aligned with the surface normal. In the case of Co deposition on the 5f-symmetry surface of i-Al-Pd-Mn, Smerdon et al. have reported that Co grows without alloying with the substrate [65]. However, they have observed no order using both scanning tunnelling microscopy and low-energy electron diffraction (LEED), until a relatively high coverage of Co is reached. For an approximately 20-ML thick Co deposit, LEED patterns show some qualitative agreement with patterns obtained from Cu films grown on the 5f-symmetry surface of i-Al-Pd-Mn. These films consist of five domains rotated by 72° with respect to each other having each a row structure at

---

[4]Body-centred cubic (bcc)-related alloy structures are generally referred to as CsCl-type expect from binary alloy structures possessing the actual CsCl structure.

Adsorbates	Substrates	Growth characteristics	References
Ag; Ag (In)	10f d-Al-Co-Ni	unordered; unordered	[50]
Ag	10f d-Al-Co-Ni	5 or $10\times$Ag(111)	[37]
Ag; Ag (In)	5f i-Al-Pd-Mn	unordered; unordered	[50]
Ag	5f i-Al-Pd-Mn	heterogeneous nucleation, 5 or $10\times$Ag(111), magic-height islands	[37, 51, 52]
Ag	2f i-Al-Pd-Mn	1 or $2\times$Ag(111)	[37]
Al	10f d-Al-Co-Ni	heterogeneous nucleation, $2\times10\times$Al(111)	[53]
Al	5f i-Al-Pd-Mn	$5\times$Al, $[01\tau] \parallel 5f$, $250-300°C$, $5\times$Al(111), $200-250°C$	[54, 55]
Al	3f i-Al-Pd-Mn	$2\times$Al(111)	[56]
Al	5f i-Al-Cu-Fe	heterogeneous nucleation	[57]
Au; Au (In)	10f d-Al-Co-Ni	$10\times$AuAl$_2$(110); $10\times$AuAl$_2$(110)	[58]
Au; Au (In)	5f i-Al-Pd-Mn	polycrystalline; ordered Au-Al	[59]
Bi	10f d-Al-Co-Ni	pseudomorphic	[60]
Bi	5f i-Al-Pd-Mn	pseudomorphic	[60]
Bi	5f i-Al-Cu-Fe	$5\times$Bi(100), magic-height islands	[52]
C_{60}	10f d-Al-Co-Ni	unordered	[61]
C_{60}	5f i-Al-Pd-Mn	heterogeneous adsorption	[62]
CD_4O	10f d-Al-Co-Ni	dissociation or low sticking probability	[63]
CD_4O	5f i-Al-Pd-Mn	unordered	[63]
C_6H_6	5f i-Al-Pd-Mn	unordered	[63, 64]
CO	10f d-Al-Co-Ni	heterogeneous adsorption	[63]
CO	5f i-Al-Pd-Mn	no adsorption	[63]
Co	10f d-Al-Co-Ni	$5\times$Co(1000)	[65]
Co	5f i-Al-Pd-Mn	$5\times$Co $5\times$Co(110)/AlCo(110)	[65] [66]
Cu	5f i-Al-Pd-Mn	$5\times$Cu, quasiperiodic row structure $5\times$Al$_4$Cu$_9$(110)	[67] [68]
Cu	2f i-Al-Pd-Mn	quasiperiodic row structure	[69]
Cu	5f i-Al-Cu-Fe	unordered	[70]
D; D_2	5f i-Al-Pd-Mn	adsorption; dissociation or no adsorption	[71]
Fe	5f i-Al-Pd-Mn	$5\times$Fe(110)/Al-Fe(110)	[72]
HCOOH	10f d-Al-Co-Ni	unordered	[63]
HCOOH	5f i-Al-Pd-Mn	dissociation	[63]
NO	10f d-Al-Co-Ni	dissociation or low sticking probability	[63]
NO	5f i-Al-Pd-Mn	dissociation	[63]
Ni	5f i-Al-Pd-Mn	$5\times$Ni(110)/Al-Ni(110)	[73]

Adsorbates	Substrates	Growth characteristics	References
O_2	5f i-Al-Pd-Mn	oxidation of Al	[74]
Pt; Pt (In)	10f d-Al-Co-Ni	polycrystalline; polycrystalline, $10 \times Al_2Pt(110)$	[50, 75]
Pt; Pt (In)	5f i-Al-Pd-Mn	unordered; unordered	[75]
S	5f i-Al-Pd-Mn	unordered	[76]
Sb	10f d-Al-Co-Ni	pseudomorphic	[60]
Sb	5f i-Al-Pd-Mn	pseudomorphic	[60]
Si	10f d-Al-Co-Ni	heterogeneous adsorption	[77]
Si	5f i-Al-Pd-Mn	heterogeneous adsorption, unordered	[78]
Sn	10f d-Al-Co-Ni	quasiperiodic	[79]
Sn	5f i-Al-Cu-Fe	pseudomorphic	[80]
Xe	10f d-Al-Co-Ni	5 or $10 \times Xe(111)$	[81]

Table 1.1: List of adsorbates deposited onto various quasicrystalline substrates. The first column lists adsorbates in alphabetic order. The notation (In) denotes that In was used as a surfactant. In the second column, the substrates are listed. 2f, 3f, 5f, and 10f are acronyms of the terms twofold, threefold, fivefold, and tenfold, respectively. Column three lists growth characteristics. Crystalline domain structures are generally indicated by the number of different domain orientations, the chemical composition, and the exposed faces parallel to the surface. Column four lists corresponding references. Only experimental results and no reviews are considered. The growth modes of Co, Ni, and Fe on the pentagonal surface of i-Al-Pd-Mn are discussed in detail in this dissertation. This list is not exhaustive.

the surface with row separations corresponding to the Fibonacci sequence[5] [67].

(c) The production of quasicrystalline coatings on crystalline substrates aims at utilising the extraordinary physical properties of quasicrystals like high hardness or low friction [36]. Al-Co-Ni [82] and Al-Cu-Fe-Cr [83] films grown on sapphire(0001) consist of quasicrystalline domains with preferred orientations related to the substrate.

Dmitrienko and Astaf'ev have demonstrated the close relationship between the i-quasicrystal and the CsCl structure by introducing a three-dimensional model for the growth of i-quasicrystals [84]. Steurer has explained the orientation of sputter-induced CsCl-type domains on d- and i-quasicrystals by the optimum matching of the average structures [85]. Widjaja and Marks have in a more general treatment employed the coincidence reciprocal-lattice-planes model with the inclusion of the interfacial energy in order to explain the orientational relationships at quasicrystal-crystal interfaces [83,86]. The few discrepancies of their results from experimentally observed in-plane relationships were attributed to the simplistic nature of the inter-

---

[5]The Fibonacci sequence, which is quasiperiodic, can be generated with the substitution rules $S \mapsto L$, $L \mapsto LS$. S and L may denote short and long distances, respectively.

action potential and/or limited kinetics present in the experimentally investigated systems resulting in metastable configurations. Discrepancies may also arise because details of the growth mode, including, e.g., nucleation, intermixing, and the formation of three-dimensional islands, have not been taken into account [55].

Nanotechnology aims at generating and using devices which consist of structures in the size range of $\sim 0.1 - 100$ nm. Two basic approaches to create such structures are top-down and bottom-up techniques [87]. In the case of top-down techniques like ultraviolet lithography [88] or direct patterning by means of scanning probe microscopy [89], a pattern is superimposed onto a substrate. In contrast, bottom-up techniques utilise self-organised growth and self-assembly in order to obtain organised atomic and molecular structures, respectively. The growth mode is governed by the competition between kinetics and thermodynamics, i.e., deposition rate and surface diffusion.

Structures of magnetic materials with dimensions in the nm range exhibit a variety of remarkable phenomena (see, e.g., Refs. [90–93]). If the size of magnetic particles is reduced below the domain wall thickness, a single-domain state is preferred [91–94]. For a further reduction of the particle size, the superparamagnetic limit is reached. Spontaneous flips of the magnetisation occur due to thermal activation [93, 94]. In addition to the reduction of the temperature, the enhancement of the anisotropy energy, via, e.g., the enhancement of the magnetocrystalline anisotropy energy, shifts the superparamagnetic limit to smaller particle sizes [94]. In the case of magnetic particles grown on substrates, magnetostriction resulting from strain present at the interface has to be additionally considered [93].

In this dissertation, the interfaces between the pentagonal surface of i-Al-Pd-Mn and evaporated Co, Ni, as well as Fe are studied. A potentially growing quasiperiodic superlattice of self-organised, discrete, crystalline domains with nm dimensions, well-defined orientations, and exhibiting ferromagnetic ordering would represent an exceptional magnetic system. Furthermore, the growth mode of crystalline structures on a quasicrystalline substrate may reveal some information on the nature of the quasicrystalline structure.

Chapter 2

Experimental setup

An i-Al-Pd-Mn quasicrystal having a nominal bulk composition of $Al_{70}Pd_{20}Mn_{10}$ was grown using the Czochralski technique. A 5f-symmetry surface with dimensions of $\sim 10 \times 8$ mm^2 was polished with diamond pastes with grain sizes down to 0.5 μm. Subsequently, the sample was mounted on a goniometer [95] and inserted into an ultrahigh-vacuum chamber having a base pressure in the lower 10^{-10} mbar range and equipped with facilities in order to perform Auger electron spectroscopy (AES), LEED, secondary-electron imaging (SEI), and longitudinal magneto-optical Kerr effect (MOKE) measurements. The sample surface was prepared by cycles of sputtering with Ar$^+$ ions (1.5 keV, 0.4 μA/mm^2) at room temperature and annealing at ~ 740 K for 40 minutes. The sample was heated from the backside by means of a resistance heater [96]. After annealing, the sample could be actively cooled down using a Cu cooling feed which was brought in thermal contact with liquid N$_2$. The temperature was measured using a chromel-alumel (K-type) thermocouple pressed onto the sample surface.

AES measurements were performed in order to determine the chemical composition in a near-surface region. AES utilises the Auger effect, in which a hole at a core level of an ionised atom is filled by an electron of lower binding energy and the available energy is transmitted in a radiationless process to the so-called Auger electron which leaves the atom with an element-specific kinetic energy [97, 98]. Auger electrons ejected from a solid provide surface-specific chemical information, since electrons with kinetic energies between 10 and 1000 eV have a mean free path in metals of typically $2 - 30$ Å [97, 99–101].

The surface structure was investigated by means of LEED and SEI. In the case of LEED, electrons with kinetic energies of $30 - 300$ eV, having a penetration depth of up to ~ 10 Å [97, 99, 100], are elastically backscattered from the surface and form diffraction patterns [97, 98]. These patterns yield information in reciprocal space, averaged over the coherence width of the incident electron beam, of long-range order and surface defects present at the surface [98, 102]. In the case of SEI [103], electrons with a kinetic energy of 2 keV, possessing a mean free path in metals of about $10 - 50$ Å [99, 100], are directed onto the sample surface. These primary electrons excite, at localised atomic positions, secondary electrons which, in a first process, subsequently scatter at the Coulomb potential of neighbouring atoms and are therefore focused in forward direction [104]. Consequently, these secondary electrons are predominantly channelled along densely packed atomic rows. In a second process, secondary-electron waves are reflected at atomic planes giving rise, due to interference between reflections at adjacent planes, to so-called Kikuchi [105] bands. As a result, SEI provides real-time information of the local atomic arrangement in a near-surface region in real space revealing the symmetry in orthographic (central) projection. Therefore, this technique ideally complements LEED studies, since it eliminates ambiguities of diffraction experiments which may introduce additional symmetries. Both techniques, LEED and SEI, were carried out with the same back-view display system [106] concentric with the sample and having a total opening angle of about 93°. Patterns were recorded using a 16-bit charge-coupled device camera [107] and were subsequently normalised by the overall response function of the display system in order to eliminate spurious signals to some extent. Patterns represented in this dissertation were obtained at normal incidence (expect for Fig. 2.1a) and the sample kept at room temperature. Fig. 2.1a presents an SEI pattern obtained from the pentagonal surface of an i-Al-Pd-Mn quasicrystal. The surface normal, which is parallel to the 5f-symmetry axis of the sample structure defined by densely packed atomic rows giving rise to the SEI patch marked with a pentagon, includes an angle of 30° with the incident electron beam which is perpendicular to the image plane. A 2f- and a 3f-symmetry axis are also marked with corresponding symbols. Kikuchi bands belonging to 5f-symmetry planes ex-

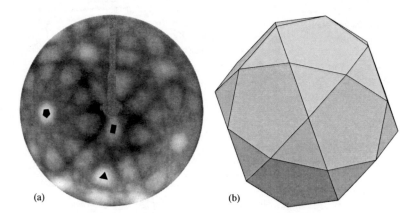

(a)   (b)

Fig. 2.1: (a) SEI pattern obtained from the 5f-symmetry surface of i-Al-Pd-Mn. The sample is tilted by a polar angle of 30° to the left. A 5f-, a 3f-, and a 2f-symmetry axis are marked with a pentagon, a triangle, and a rectangle, respectively. The marked 5f-symmetry axis corresponds to the surface normal. The marked 2f-symmetry axis, including an angle of ∼ 31.72° with the adjacent 5f-symmetry axes, is the intersection line of two 5f-symmetry planes giving rise to Kikuchi bands. Bands belonging to 2f-symmetry planes include SEI patches related to 2f-, 3f-, as well as 5f-symmetry axes. The shadow of the electron gun and its holder is visible running vertically from the top towards the centre of the pattern. (b) An icosidodecahedron possessing i-symmetry and oriented in accordance with the orientation of the sample giving rise to the pattern shown in (a).

clusively contain SEI patches related to 2f-symmetry axes, while bands belonging to 2f-symmetry planes include patches related to 2f-, 3f-, as well as 5f-symmetry axes. In Fig. 2.1b, an icosidodecahedron, possessing i-symmetry, is illustrated. It is oriented in accordance with the orientation of the sample which gives rise to the SEI pattern represented in Fig. 2.1a and demonstrates that the symmetry of the atomic arrangement, averaged both over the spot size of the electron beam and a depth given by the mean free path of 2-keV electrons, is instantaneously identifiable using SEI.

When reflected from a magnetised surface, the plane of polarisation of linearly polarised light is rotated in proportion to the magnetisation due to the coupling between the electrical field of the light and the electron spins through a spin-orbit interaction [108]. Utilising this so-called MOKE, measurements in the longitudinal setup, in which the magnetic field is applied parallel to the surface and in the plane of incidence of the light, were performed simultaneously with the deposition of Co, Ni, and Fe in order to detect in-plane magnetic ordering as a function of film thickness. For this purpose, a linearly polarised, intensity-stabilised He-Ne laser [109] (wavelength of 632.8 nm) along with a photoelastic modulator [110] was used as a light source and the alternating current component of the magnetic signal parallel to the applied magnetic field was measured with a phase-sensitive detector. The angle of incidence was $45°$.

In the case of Ni evaporation, scanning electron microscopy with polarisation analysis (SEMPA) of secondary electrons [91] was additionally carried out in a different ultrahigh-vacuum chamber. A 5f-symmetry surface of an additional i-Al-Pd-Mn quasicrystal, grown using the Bridgman technique and having a nominal bulk composition of $Al_{70}Pd_{20}Mn_{10}$, was also prepared with Ar^+-ions sputtering and annealing. The sample quality was surveyed using LEED and AES. Performing SEMPA, the sample surface is scanned with a focused beam of electrons with a typical kinetic energy of $2 - 10$ keV. The amount of secondary electrons emitted from the surface depends on both the local chemical composition and topography. By means of a Mott detector, the spin polarisation of these electrons is detected which is in a first-order approximation proportional to the magnetisation of the surface [111].

Co, Ni, and Fe of 99.99+%, 99.99+%, and 99.999% purity, respectively, were evaporated onto the sample surface using a power-regulated atomic-beam source. In the cases of Co and Ni, the sample was kept at room temperature during deposition. In the case of Fe, evaporation on the sample kept both at room temperature as well as at ~ 340 K was carried out. The deposition rates of Co, Ni, and Fe, which were in the range of $0.5 - 1$ Å/minute, were repeatedly calibrated by measuring the L_3MM Auger signals of Cu, Cu, and Fe, respectively, during the deposition on polycrystalline Cu. As a most reliable value for the mean free path of the electrons in Co, 4 Å was chosen [101]. The mean free path of the electrons in Ni and Fe was calculated using the methods of Tanuma et al. [99] as well as Seah and Dench [100]. Some systematic deviations from the determined values of the deposition rate cannot be excluded.

Chapter 3

Co deposition on the pentagonal surface of icosahedral Al-Pd-Mn

3.1 The clean substrate surface

After repeatedly annealing the sputtered sample at 740 K for 40 minutes, a LEED pattern as shown in Fig. 3.1a is obtained from the pentagonal surface at a primary-electron energy of 105 eV. The quality of the pattern indicates a well-established quasicrystalline order present at the surface [98, 102]. At the top right-hand side of the pattern, basis vectors e_i are plotted in accordance with those introduced by Schaub et al. [42] in order to index LEED spots. Table 3.1 lists intense spots with numbers referring to those marked in Fig. 3.1a, corresponding indices, the experimentally determined components of the scattering vectors parallel to the surface $k_\parallel$, as well as resultant lattice parameters a ($= 2\pi/|k_\parallel|$). Analysing LEED patterns recorded at various primary-electron energies yields a length of the basis vectors e_i of 0.6317 ± 0.0118 Å^{-1}. This value agrees with 0.6278 ± 0.0085 Å^{-1} found by Schaub et al. [42]. In contrast, Shen et al. have reported a value of ~ 0.671 Å^{-1} which is considerably higher, in particular with respect to the bulk value of 0.616 Å^{-1} [112].

3.2 Growth of AlCo and Co domains

The LEED pattern presented in Fig. 3.1b was obtained under the same conditions as the pattern shown in Fig. 3.1a but with a Co deposit of 0.34-ML thickness[1], evaporated onto the sample kept at room temperature. Although having much less intensity, (21001)- and (42013)-type LEED spots are still detectable. Other spots are hardly [(32002)-type] or not visible anymore. Considering a mean free path of a few Å for the low-energy electrons [97, 99–101], the disappearance of most LEED spots associated with the quasicrystalline surface structure after the deposition of only 0.34 ML of Co suggests a partial loss of long-range order in an initially quasicrystalline surface layer. Since, in accordance with Smerdon et

---

[1]1 ML equals a film thickness of 2.04 Å, considering structure and orientation of the growing film. CsCl-type AlCo as well as bcc Co domains have lattice constants of ~ 2.88 Å and expose their (110) faces.

al. [65], diffusion of Co atoms into Al-Pd-Mn was not detectable with the AES measurements, this loss is not a result of intermixing of Co with the substrate but of atomic rearrangements caused by adsorbed Co atoms.

Patterns which were obtained from the pentagonal surface covered with Co deposits of 0.67 and 1.34 ML at a primary-electron energy of 105 eV are presented in Figs. 3.1c and d. Together with patterns obtained for the same deposits but recorded

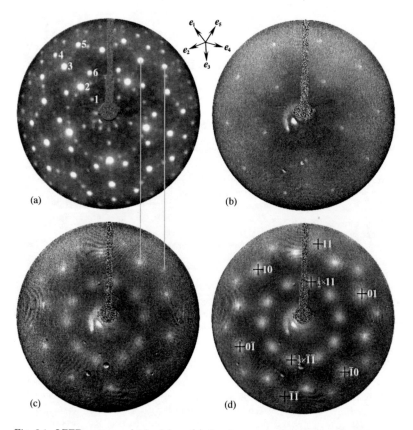

Fig. 3.1: LEED patterns obtained from (a) the clean pentagonal surface of i-Al-Pd-Mn and Co deposits of (b) 0.34, (c) 0.67, and (d) 1.34 ML at a primary-electron energy of 105 eV. At the top right-hand side of pattern (a), basis vectors e_i are plotted which are used in order to index diffraction spots numbered on the right-hand side in the pattern in accordance with Table 3.1. Intense spots appear to have relatively large dimensions due to saturation of the detection system. In pattern (d), spots originating from one domain are marked with black crosses and indexed according to the scheme described in the text. The grey lines indicate that 01- and 11-type spots of the domains are found at the same positions as (32002)- and (42013)-type spots of the substrate.

Table 3.1: Selection of LEED spots found in Fig. 3.1a. The numbers in the first column correspond to those inscribed in the pattern. In the second, third, and fourth column, corresponding vectors $(n_1 n_2 n_3 n_4 n_5)$, components of the scattering vectors parallel to the surface $\boldsymbol{k}_\parallel$, and resultant lattice parameters a are listed, respectively.

#	$(n_1 n_2 n_3 n_4 n_5)$	$|\boldsymbol{k}_\parallel|$ [Å^{-1}]	a [Å]
1	(11001)	1.022	6.147
2	(21001)	1.654	3.799
3	(32002)	2.676	2.348
4	(42002)	3.308	1.899
5	(42013)	3.146	1.997
6	(21002)	1.944	3.232

at different primary-electron energies, they reveal an arrangement of domains each possessing a structure derived from a CsCl-type (110) surface. These domains are rotated by 72° with respect to each other. In Fig. 3.1d, spots originating from one domain are marked with black crosses and indexed with the components of reciprocal surface-lattice vectors. 11 and 1$\bar{1}$ vectors correspond to $\frac{1}{2}s \times [\bar{1}10]$ and $\frac{1}{2}s \times [001]$ direction vectors found in the CsCl structure, respectively, s being the lattice constant. The 11-type spots are second-order diffraction spots accompanying the innermost spots, indexed $\frac{1}{2} \times 11$ and $\frac{1}{2} \times 1\bar{1}$, corresponding to $s \times [\bar{1}10]$ direction vectors. The grey lines indicate that 01- and 11-type LEED spots of the domains match, within the accuracy of the measurement, with (32002)- and (42013)-type spots obtained from the quasicrystalline surface structure. Hence, taking the lattice parameter of 2.348 Å which corresponds to the (32002)-LEED spot (Table 3.1), s is approximately 2.88 Å [≈ 2.348 Å$/sin(arctan\sqrt{2})$]. Since the angle between the 0$\bar{1}$ and 10 reciprocal vectors is $arccos\frac{1}{3} \approx 70.53°$ but domains are rotated by 72° with respect to each other, the 10-type spots found in the patterns effectively consist of two spots separated by an azimuthal angle of $\sim 1.47°$. However, this separation of spots is not discernible in the LEED patterns due to the limited resolution and the broadness of individual spots which depends on the small lateral size of the domains. Spot-profile analyses [44,113] reveal domain diameters of ~ 20 Å for film thicknesses of 0.67, 1.34, and 2.02 ML.

Both structure and lattice constant suggest that for a Co deposit of 0.67 ML AlCo-type domains are formed on the 5f-symmetry surface of i-Al-Pd-Mn, since AlCo possesses the CsCl structure with a lattice constant of ~ 2.86 Å [114,115], while other Al-Co alloys exhibit different structures [115]. A contribution from the quasicrystal structure to the LEED spot intensities observed after the deposition of 0.67 and 1.34 ML can be excluded since patterns show, different from Fig. 3.1a, a 10f rotational symmetry for the entire range of the primary-electron energy used here. Therefore, the formation of islands on the quasicrystalline substrate can be ruled out. Furthermore, considering the mean free path of low-energy electrons [97, 99–101], the quasicrystalline order is vanished at least in a depth of approximately 2 Å at this stage of growth.

The formation of AlCo domains for the initial, submonolayer Co deposition is comprehensible due to the chemical composition found at the pentagonal surface of i-Al-Pd-Mn. Carrying out dynamical LEED analysis, Gierer et al. have found that the outermost atomic layer is primarily composed of Al followed by a layer containing about 50% Al and 50% Pd atoms [116]. This finding has been confirmed by

low-energy ion-scattering measurements performed by Bastasz [117]. The formation
of 3 and 5 ML thick AlCo layers has been observed as a result of Co deposition on
Al(100) and Al(110), respectively, followed by the growth of pure Co layers [118].
In the present case, no intermixing of Co and Al normal to the surface is observed
at the interface which, however, does not rule out the formation of an atomic surface layer consisting of AlCo domains formed for a submonolayer deposit of Co.
Potentially, this well-ordered surface alloy together with the limited amount of Al
present at the substrate surface prevents intermixing.

The contrast quality in the pattern shown in Fig. 3.1d is superior to the pattern
of Fig. 3.1c. This indicates a more distinct surface structure. Since no diffusion
normal to the surface occurs and similar patterns like that presented in Fig. 3.1d
are also obtained for Co deposits of 2.02 and 2.69 ML, it follows that the onset
of the growth of bcc Co domains is observed at this stage of growth. Co films in
the metastable bcc structure have been reported first by Prinz [119]. They were
produced by atomic-beam epitaxy on GaAs(110) with thicknesses up to 357 Å and
a lattice constant of 2.827 Å. Among others, bcc Co films have also been grown
epitaxially on AlFe(100) [120] and Fe(100) [121]. In the case of the deposition on
AlFe(100), however, $5 - 10$ at.% Al was present in the Co film. On Fe(100), bcc
Co(100) films with an out-of-plane lattice parameter of 2.87 ± 0.02 Å were found
for thicknesses up to 20 Å. According to these findings, the growth of bcc Co on an
appropriate substrate fulfilling conditions for epitaxy is well feasible.

Before clarifying to what extent epitaxial conditions are fulfilled for the growth
of AlCo and bcc Co domains on the pentagonal surface of i-Al-Pd-Mn and thereby
explaining how an atomic surface layer consisting of CsCl-type domains may be
stabilised, the processes involved in the formation of the interface are summarised
as follows. At first, the deposition of Co results in the loss of long-range order in
an initially quasicrystalline surface region. Subsequently, a layer is formed which
consists of domains possessing a structure derived from a CsCl-type (110) surface.
On these AlCo domains, bcc Co grows up to a thickness of ~ 3 ML. Both AES
measurements and thickness-independent domain sizes suggest a uniform covering
of the substrate.

3.3   AlPd domains induced by sputtering

Five CsCl-type AlPd domains exposing their (113) faces[2] and rotated by 72° with
respect to each other are observed at the surface using SEI after removing Co deposits of less than ~ 1 ML by means of sputtering as well as after sputtering the
clean surface (Fig. 3.2a, cp. Refs. [45, 46]). If Co deposits of more than ~ 1 ML
are removed, however, the surface consists of CsCl-type AlPd(110) domains rotated by 72° increments (Fig. 3.2b, cp. Refs. [46–49]). Naturally, during completely
removing Co by means of sputtering, relative concentrations of Al, Pd, and Mn
which are also removed from the surface cannot be controlled. However, once either one or the other alignment of domains is established it was found to be stable

---

[2]In the following, AlPd domains exposing their (113) and (110) faces are referred to as
AlPd(113) and AlPd(110) domains, respectively.

against sputtering at room temperature and normal incidence, as performed in this investigation.

As already mentioned in the introduction, sputtering the clean pentagonal surface of i-Al-Pd-Mn leads to a modified chemical composition at the surface which destabilises the quasicrystalline structure. As a result, both CsCl-type AlPd(110) and AlPd(113) domains have been observed [45–49]. Single-domain overlayers [45–47], breaking the symmetry of the substrate, as well as overlayers consisting of five domains rotated by 72° with respect to each other [45, 48], conserving the substrate symmetry on a global scale, have been found in both cases. Naumović et al. have reported that single-domains are generated due to anisotropic sputtering [45]. These findings indicate that the mechanism relevant for the formation of one or the other alignment of domains is very sensitive to, at least, sputtering conditions. In both cases, the quasicrystalline surface structure is restored by annealing as a result of the re-established chemical composition caused by diffusion.

The reason for the observation of two orientations of AlPd domains at the sputtered Al-Pd-Mn surface is twofold. (a) The domain orientation with the [110] direction parallel to a 5f-symmetry axis and the [1$\bar{1}$0] direction aligned with a 2f-symmetry axis of the i-substrate is the optimum matching of the average structures [18, 85]. This orientational relationship is illustrated in Fig. 3.3. It allows, in accordance with the number of 5f-symmetry axes of the i-structure, six different alignments with respect to a 5f-symmetry axis. With reference to Fig. 3.3, these alignments have their [110], [1 1 $\overline{2.83}$], [3.32 $\overline{1.03}$ $\bar{1}$], [$\overline{1.03}$ 3.32 $\bar{1}$], [1 $\overline{12.39}$ 13.03], and [$\overline{12.39}$ 1 $\overline{13.03}$] directions parallel to 5f-symmetry axes (non-integers are rounded to two decimals). However, since the (1$\bar{1}$0) plane, which is equal to a 2f-symmetry

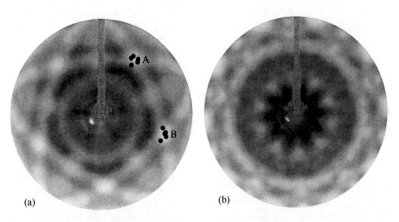

(a)   (b)

Fig. 3.2: SEI patterns obtained after (a) sputtering the clean surface or removing Co deposits of less than $\sim$ 1 ML and (b) removing Co deposits of more than $\sim$ 1 ML. The assemblies of dots labelled A and B in the pattern (a) correspond to the superposition of directions, resulting from the rotation of domains by 72° increments, in which secondary electrons would predominantly arrive at the display screen and generate the corners of a pentagon in the cases of the [1 1 $\overline{2.83}$] and the [3.32 $\overline{1.03}$ $\bar{1}$] alignment, respectively.

plane of the quasicrystal, is a mirror plane for both the cubic and the i-structure, the $[3.32\,\overline{1.03}\,\overline{1}]$ and $[\overline{1.03}\,3.32\,\overline{1}]$ as well as the $[1\,\overline{12.39}\,\overline{13.03}]$ and $[\overline{12.39}\,1\,\overline{13.03}]$ alignments are identical (in the following referred to as $[3.32\,\overline{1.03}\,\overline{1}]$ and $[1\,\overline{12.39}\,\overline{13.03}]$ alignments). (b) The four remaining alignments are equal with respect to the substrate symmetry but differ, for the surface normal parallel to a 5f-symmetry axis of the substrate, with respect to faces exposed to the surface and possible domain-domain interfaces. These differences may partly determine the occurrence or non-occurrence of alignments, since surface energies responsible for the stability of a surface are orientation-dependent [122] and dissimilar misalignments at unequal domain-domain interfaces induce different strains at the interface. However, responsible for the occurrence of selected alignments is also a mechanism by which the direction of the incident Ar^+-ion beam during sputtering relative to the sample orientation decides on the resulting domain orientations (see, e.g., Ref. [123]). As experimentally observed for ion-beam-assisted deposition of different metal films [124], domains are not uniformly damaged by sputtering according to their orientations with respect to the incident ion beam since they possess channelling directions, i.e., crystallographic directions along which ions predominantly penetrate into the solid. Domains which are less damaged expand by a recrystallisation process.

While the observed AlPd(110)-domain overlayers correspond to the [110] alignment, the overlayers consisting of nominal AlPd(113) domains may correspond to the $[1\,1\,\overline{2.83}]$ as well as the $[3.32\,\overline{1.03}\,\overline{1}]$ alignment. These alignments are very similar regarding their orientation with respect to the substrate symmetry. For the $[1\,1\,\overline{2.83}]$ alignment, the $[11\overline{3}]$ direction differs from the adjacent 5f-symmetry axis by only $\sim 1.33°$. Similarly, the $[3\overline{1}\overline{1}]$ direction and the adjacent 5f-symmetry axis span an angle of $\sim 1.87°$ in the case of the $[3.32\,\overline{1.03}\,\overline{1}]$ alignment (Fig. 3.3). For a nominal AlPd(113) single-domain overlayer the patches observable in SEI form the corners of an almost equilateral pentagon (see Fig. 2c in Ref. [46]) and are, in the case of the $[1\,1\,\overline{2.83}]$ alignment, related to atomic rows in $[11\overline{1}]$, $[01\overline{1}]$, $[10\overline{1}]$, $[\overline{1}1\overline{4}]$, and $[1\overline{1}4]$ directions. This situation is illustrated in orthographic projection in Fig. 3.4. The angle between the $[11\overline{3}]$ and $[11\overline{1}]$ directions is $29.50°$. The $[01\overline{1}]$, $[10\overline{1}]$, $[\overline{1}1\overline{4}]$, and $[1\overline{1}4]$ directions each span an angle of $31.48°$ with the $[11\overline{3}]$ direction. In the plane perpendicular to the $[11\overline{3}]$ direction, the angles between $[11\overline{1}]$ and $[01\overline{1}]$, $[01\overline{1}]$ and $[\overline{1}1\overline{4}]$, as well as $[\overline{1}1\overline{4}]$ and $[1\overline{1}4]$ directions are 73.22, 67.11, and $79.33°$, respectively. The $[01\overline{1}]$ and $[10\overline{1}]$ directions almost match with 2f-symmetry axes of the i-structure (misaligned by $1.83°$), while the $[1\overline{1}0]$ direction is aligned with a 2f-symmetry axis. In the case of a nominal AlPd(113) domain related to the $[3.32\,\overline{1.03}\,\overline{1}]$ alignment, this direction corresponds to one of the corners of the pentagon observable in SEI. The four other corners correspond to atomic rows in $[1\overline{1}\overline{1}]$, $[10\overline{1}]$, $[4\overline{1}1]$, and $[41\overline{1}]$ directions. Unfortunately, the limited resolution of SEI and the rotation of domains by $72°$ increments, resulting in the superposition of patches, do not allow to distinguish, by means of SEI, between the $[1\,1\,\overline{2.83}]$ and $[3.32\,\overline{1.03}\,\overline{1}]$ alignments. This situation is represented in Fig. 3.2a. The assembly of dots labelled with A (B) corresponds to the superposition of directions in which secondary electrons would generate the corners of a pentagon in SEI for the $[1\,1\,\overline{2.83}]$ ($[3.32\,\overline{1.03}\,\overline{1}]$) alignment. Accordingly, AlPd domains in the [110] as well as potentially both the $[1\,1\,\overline{2.83}]$ and $[3.32\,\overline{1.03}\,\overline{1}]$ alignments, but not in the

[1 $\overline{12.39}$ $\overline{13.03}$] alignment, are observed. This suggests that the two factors determining the occurrence of alignments, the alignment-selecting sputtering mechanism and orientation-dependent surface as well as domain-domain interface alignments, only allow the three first alignments and sensitively induce the formation of either AlPd(110) or nominal AlPd(113) domains.

Fig. 3.5a shows a LEED pattern obtained at a primary-electron energy of 70 eV from the AlPd(110) domains after annealing at $\sim$ 520 K for 30 minutes. Spots originating from one of the five domains are marked with black crosses and indexed in accordance with Fig. 3.1d. Accordingly, 11 and 1$\overline{1}$ reciprocal surface-lattice vectors correspond to $\frac{1}{2}t \times [\overline{1}10]$ and $\frac{1}{2}t \times [001]$ direction vectors of the CsCl structure,

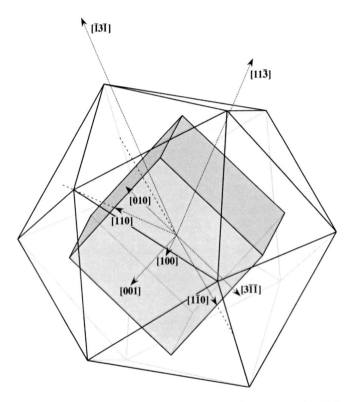

Fig. 3.3: Illustration of the orientational relationship between the i-structure of the substrate, represented by an icosahedron, and the cubic structure of the sputter-induced domains, represented by a cube. The vectors [100], [010], and [001] point from the centre of the cube, which is the same as the centre of the icosahedron, to the centres of three faces of the cube. The [110] direction is parallel to a 5f-symmetry axis of the icosahedron and the [1$\overline{1}$0] direction is parallel to a 2f-symmetry axis.

Fig. 3.4: Orthographic projection of directions found in the cubic structure. The two circles represent polar angles of 20 and 40°. The $[11\bar{1}]$, $[01\bar{1}]$, $[\bar{1}14]$, $[1\bar{1}4]$, and $[10\bar{1}]$ directions form the corners of an almost equilateral pentagon. The $[11\bar{3}]$ direction is perpendicular to the image plane. The angle between the $[11\bar{3}]$ and $[11\bar{1}]$ directions is 29.50°. The $[01\bar{1}]$, $[10\bar{1}]$, $[\bar{1}14]$, and $[1\bar{1}4]$ directions each span an angle of 31.48° with the $[11\bar{3}]$ direction. Azimuthal angles in the plane perpendicular to the $[11\bar{3}]$ direction between adjacent directions are inscribed in grey.

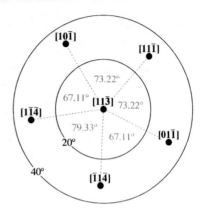

$t \approx 2.88$ Å being the lattice constant. The 11-type spots are, due to second-order diffraction, belonging to the fundamental innermost spots visible in the pattern (not labelled). First-order spots of $1\bar{1}$-type are also observable, labelled $\frac{1}{2} \times \bar{1}1$ and $\frac{1}{2} \times 1\bar{1}$. The domains are tilted by approximately $0.5 - 1°$ relative to the surface normal in directions corresponding to that indicated with an arrow at the top right-hand side of the pattern for the labelled domain. With respect to the i-symmetry, the tilt proceeds in 2f-symmetry planes away from the 5f-symmetry axis being perpendicular to the surface towards adjacent 3f-symmetry axes (see Fig. 3.3). Due to the tilt, neighbouring 11-type spots, for instance, are alternately separated in azimuth by more and less than 36° and the pattern exhibits 5f rather than 10f rotational symmetry. The tilt is also observable in the SEI pattern in Fig. 3.2b. Five domains with the [110] direction, a 2f-symmetry axis, exactly parallel to the 5f-symmetry axis of the substrate, parallel to the surface normal, and rotated by 72° with respect to each other would generate a pattern with 10f rotational symmetry. In contrast, the pattern in Fig. 3.2b exhibits a 5f rotational symmetry. Unfortunately, the LEED patterns obtained from the surface consisting of nominal AlPd(113) domains are of poor quality, neither revealing the surface structure, clarifying the occurrence of the $[1\,1\,\overline{2.83}]$ and/or $[3.32\,\overline{1.03}\,\overline{1}]$ alignments, nor displaying a possible tilt.

Regarding the tilt of AlPd(110) domains, it is conceivable that the domain orientation is influenced by domain boundaries, since domains are rather small and therefore have relatively large boundary areas compared to their volume. Spot-profile analyses suggest domain diameters of ~ 30 Å. The thickness of the AlPd layer is at least $20 - 30$ Å, which is the information depth of SEI [99, 100], but is limited by the penetration depth of some tens of Å for the ions used for sputtering [125].

Co domains, which expose their (110) faces, show no tilt, at least for thicknesses up to ~ 3 ML. This is apparent from the pattern in Fig. 3.5b obtained from a deposit of 1.34 ML of Co since it exhibits 10f rotational symmetry. Like the pattern in Fig. 3.5a, it was recorded at a primary-electron energy of 70 eV.

In summary, one arrives at the following conclusions. (a) The formation of AlPd on i-Al-Pd-Mn with a 5f-symmetry plane parallel to the surface is caused by the modified chemical composition as a result of sputtering. (b) In order to conserve the substrate symmetry (provided that sputtering is isotropic), the resulting AlPd layer has to consist of $5n$ domains, whereas $n \in \mathbf{N}$ and domains within one set are azimuthally rotated by $72°$ with respect to each other. (c) The fundamental orientation of these domains with respect to the substrate is governed by the optimum matching of the average structures allowing four different types of alignments (therefore, $n \in \{1, 2, 3, 4\}$). (d) Due to an alignment-selecting sputtering mechanism and different surface as well as domain-domain interface alignments, not all alignments exist and the fundamental orientation of each existing alignment may slightly be modified.

3.4 Quasi-epitaxial growth

The deposition of only 0.34 ML of Co on the pentagonal surface of i-Al-Pd-Mn results in the loss of long-range order in an initially quasicrystalline surface region. A Co deposit of already 0.67 ML distinctly exhibits a texture of AlCo domains exposing their CsCl-type (110) faces parallel to the surface. These findings suggest that the interface between the i- and the cubic structure is already formed by atomic rearrangements in an initially quasicrystalline surface region at this stage of growth. This is conceivable, since, due to the optimum matching of the discrete average structure of the substrate and the CsCl structure of the growing domains, having a similar lattice constant like AlPd, only relatively minor atomic displacements are expected in order to form the interface. Therefore, this growth mode may be referred to as quasi-epitaxial. The optimum matching is immediately apparent in

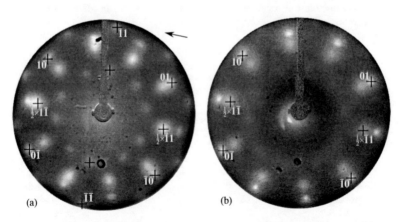

Fig. 3.5: LEED patterns obtained at a primary-electron energy of 70 eV from (a) the AlPd(110) domains after annealing at ~ 520 K for 30 minutes and (b) a Co deposit of 1.34 ML on the 5f-symmetry surface of i-Al-Pd-Mn. In both patterns, spots originating from one of the five domains are marked with black crosses and indexed according to Fig. 3.1d. The arrow at the top right-hand side of pattern (a) indicates the tilt direction of the labelled domain by about $0.5 - 1°$ away from the surface normal.

the LEED patterns presented in Fig. 3.1. (32002)- and (42013)-type spots obtained from the quasicrystalline surface structure match with 01- and 11-type spots of the crystalline structure. Co in the bcc structure grows epitaxially on the AlCo domains. In contrast to AlPd domains formed by means of sputtering the pentagonal surface of i-Al-Pd-Mn, only one alignment obeying the optimum matching of the average structures of adsorbate and substrate is found. This alignment has to be favoured due to alignment-dependent surface and domain-domain interface energies.

Although the interface between the quasicrystalline and the crystalline structure is already formed for a Co deposit of 0.67 ML, nominal AlPd(113) domains are still observed after removing Co by means of sputtering, as it is also the case after removing smaller Co deposits or sputtering the clean pentagonal surface. In contrast, AlPd(110) domains are found after the removal of deposits of more than ~ 1 ML. The interface is, compared to deposits of less than ~ 1 ML, more extensively developed. Performing sputtering, the established structural CsCl(110) character of the interface favours the formation of AlPd(110) domains.

3.5 Row structure at the surface of Co domains

Five Co domains exposing their (110) faces and possessing the bcc structure epitaxially grow up to a Co deposit of ~ 3 ML on five CsCl-type AlCo domains, which have previously formed and also expose their (110) faces, as a result of Co evaporation onto the pentagonal surface of i-Al-Pd-Mn (Section 3.2). For larger Co deposits, the surface structure is modified. This is apparent in Fig. 3.6 which presents LEED patterns obtained from Co deposits of 2.69, 3.36, 4.70, and 13.44 ML at a primary-electron energy of 105 eV. Compared to pattern (d) shown in Fig. 3.1, which was obtained from a deposit of 1.34 ML, $\frac{1}{2} \times 11$-type LEED spots are hardly discernible for a 2.69-ML deposit and have disappeared for a 3.36-ML deposit (regarding the indexing scheme see Section 3.2). Similarly, $\frac{1}{2} \times 1\bar{1}$-type spots, which are observable for a deposit of, e.g., 1.34 ML at different primary-electron energies, disappear at this stage of growth. For a film thickness of 3.36 ML, streaks become discernible which are more distinct for larger Co deposits. Parallel streaks indicate long-range ordering in one direction related to a period of ~ 2.5 Å. This value and the orientation of streaks correspond to the interatomic distance found in the $[\bar{1}11]$ as well as $[\bar{1}1\bar{1}]$ directions in the bcc structure of the Co domains exposing their (110) faces. Since $\frac{1}{2} \times 11$- and $\frac{1}{2} \times 1\bar{1}$-type spots are not observable anymore, $s \times [\bar{1}10]$ and $s \times [001]$ interatomic directions are not present in the formed surface structure which consequently consists of one-dimensional atomic rows parallel to $[\bar{1}11]$ and $[\bar{1}1\bar{1}]$ directions. The width of the streaks suggests [44, 113] an average length of rows of ~ 15 Å.

In the case of a Ag(111) film grown on GaAs(110), Smith et al. observed that atomic rows are formed at the surface which possess a periodic atomic structure and are quasiperiodically arranged [126]. Consequently, the LEED patterns of this surface structure show periodically spaced streaks perpendicular to the rows, corresponding to the periodic atomic structure, and spots at positions of linear combinations of the incommensurate reciprocal-space periods of the quasiperiodic arrangement of rows within the streaks [127]. Ledieu et al. have found a similar row structure at the surface of Cu domains which have been grown on the 5f-symmetry surface of i-Al-Pd-Mn [67]. In the present case, streaks visible in the LEED patterns exhibit no fine structure consisting of spots. However, streaks show an intensity modulation indicating a certain, yet undetermined arrangement of rows.

On a Co domain exposing the (110) face, rows in both the $[\bar{1}11]$ and the $[\bar{1}1\bar{1}]$ direction may equally be formed. As a result, five times two domains which exhibit a row structure and are rotated by $\sim 1.47°$ with respect to each other exist since the $[\bar{1}11]$ and the $[\bar{1}1\bar{1}]$ directions span an angle of $arccos\frac{1}{3} \approx 70.53°$ but Co domains are rotated by $72°$ with respect to each other. The rotational misalignment of $\sim 1.47°$ is not visible in the patterns due to the limited resolution and the broadness of the streaks. The feigned 10- and 11-type spots visible in the patterns in Figs. 3.6c and d

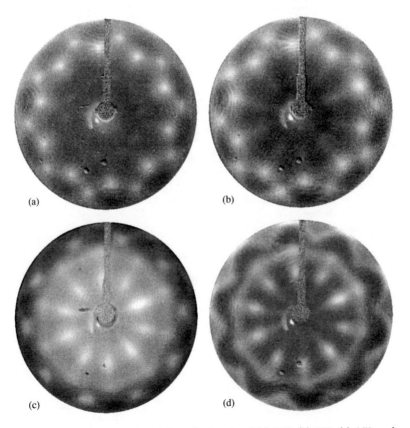

Fig. 3.6: LEED patterns obtained from Co deposits of (a) 2.69, (b) 3.36, (c) 4.70, and (d) 13.44 ML at a primary-electron energy of 105 eV.

Fig. 3.7: Dependence of saturation magnetisation and Co deposit. The line is a fit
of the data points recorded for deposits between 1.5 and 6.5 ML. In the inset, MOKE
signals obtained from the clean substrate
as well as Co deposits of about 4.3, 9.3, and
18.5 ML are shown. The axis corresponding to the magnetisation (applied magnetic
field) runs in vertical (horizontal) direction
and the point of origin is marked with a
cross. The coercive field is about 200 Oe
($\approx$ 15915 A/m) for all measured Co deposits.

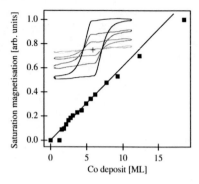

are actually intersection points of streaks. In pattern (b), actual spots may still
contribute to the pattern formation.

3.6   Magnetic ordering

In-plane magnetic ordering was investigated by means of longitudinal MOKE measurements for Co deposits up to $\sim$ 20 ML. Fig. 3.7 illustrates the thickness dependence of the saturation magnetisation, i.e., the MOKE intensity. MOKE signals as
a function of the applied magnetic field obtained from the clean substrate as well
as deposits of 4.3, 9.3, and 18.5 ML are shown in the inset. The magnetic field
was applied with respect to the orientation of the sample identifiable in Fig. 3.1
along the horizontal, i.e., the [001] directions found in the AlCo as well as Co domains, which expose their (110) faces, span angles of about 13.3, 49.3, 85.3, 121.3,
and 157.3° with the applied field. The line in Fig. 3.7 fits data points recorded for
deposits between 1.5, the onset of ferromagnetism, and 6.5 ML. For thin ferromagnetic films ($\lesssim$ 50 Å), the MOKE signal depends linearly on the film thickness if the
magnetisation of individual layers is thickness independent [128]. Therefore, the
line suggests that no magnetically dead layer is present since it approximately runs
through the point of origin. The entire Co deposit is ferromagnetic for sufficiently
thick films ($\gtrsim$ 1.5 ML). This indicates in accordance with the observations made by
means of AES and LEED (Section 3.2) that no intermixing which results in non-
ferromagnetic layers is present at the interface. The MOKE intensities measured
for deposits of more than about $7 - 8$ ML exhibit, as a function of film thickness,
a smaller increase compared to smaller deposits which may be due to structural
changes in the film.

Chapter 4

Ni deposition on the pentagonal surface of icosahedral Al-Pd-Mn

4.1 Experimental results

After sputtering the 5f-symmetry surface of i-Al-Pd-Mn at room temperature and normal incidence, an overlayer is observed by means of SEI which consists of five AlPd(113) domains azimuthally rotated by 72° with respect to each other (cp. Fig. 4.1a with Fig. 3.2a). An SEI and a LEED pattern obtained from the pentagonal surface after annealing the sample at $\sim$ 740 K for 40 minutes are shown in Figs. 4.1b and c, respectively. LEED spots are numbered according to Fig. 3.1a. Corresponding indices (indexing scheme explained in Section 3.1), experimentally determined components of the scattering vectors parallel to the surface, and resultant lattice parameters are listed in Table 3.1.

Fig. 4.2 presents Auger signal intensities as a function of the Ni deposit. The intensities of the Mn $L_3M_{2,3}M_{4,5}$ Auger signal obtained from the bare substrate and for a Ni deposit of 0.3-ML thickness[1] are comparable. In contrast, the attenuation of the intensities of the Pd $M_4N_{4,5}N_{4,5}$ signal as a function of the Ni deposit shows an exponential behaviour for the entire deposition process. Intensities of the Ni LMM signals exhibit a steady increase only for deposits of more than $\sim$ 0.6 ML. Due to the overlap of the Al LMM and Ni MNN Auger signals, Al signal intensities cannot reliably be analysed. However, Pd and Mn signals assume half of their initial values for a Ni deposit of $\sim$ 2 ML, while intensities of the Ni signal increase to half of their saturation values for a deposit of $\sim$ 3.5 ML.

LEED spots obtained from the quasicrystalline structure are observable with decreasing intensities up to Ni deposits of $\sim$ 1.5 ML. Additional spots originating from the growing film are first observed for a 1-ML deposit. Fig. 4.3a shows a LEED pattern obtained from the corresponding structure at a primary-electron energy of 115 eV after evaporating 1.9 ML of Ni. Together with patterns recorded at different primary-electron energies, it reveals an arrangement of five domains each possessing a surface structure derived from a CsCl-type (110) surface and rotated by 72° with respect to each other. Spots originating from one domain are marked with black

---

[1]1 ML $\approx$ 2.04 Å, considering structure and orientation of the growing film. CsCl-type Al-Ni and bcc Ni domains have a lattice constant of $\sim$ 2.88 Å and expose their (110) faces.

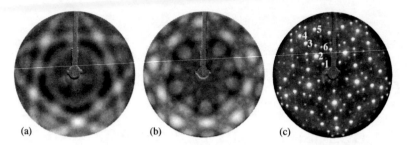

(a)   (b)   (c)

Fig. 4.1: SEI patterns recorded after (a) sputtering the pentagonal surface of i-Al-Pd-Mn and (b) subsequent annealing at ∼ 740 K for 40 minutes. (c) LEED pattern obtained from the same surface as in (b) at a primary-electron energy of 115 eV. Spots are numbered on the right-hand side in accordance with Fig. 3.1a.

Fig. 4.2: Pd, Ni, and Mn Auger signal intensities as a function of the Ni deposit. The intensities of the Mn $L_3M_{2,3}M_{4,5}$ signal obtained from the bare substrate and after the deposition of 0.3 ML of Ni (second data point) are comparable. The value for the Ni $L_3M_{4,5}M_{4,5}$ signal obtained at a Ni deposit of 0.3 ML (first data point) is an upper limit and is by a factor of ∼ 2.5 lower than the value obtained for a 0.6-ML deposit (second data point). The Pd $M_4N_{4,5}N_{4,5}$ signal exhibits an exponential decrease. With respect to the original data, the Auger signal intensities of Pd and Mn were multiplied by 1/3 and 2, respectively.

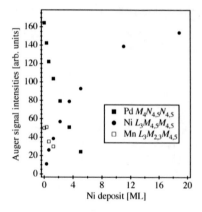

crosses and indexed with the components of corresponding reciprocal surface-lattice vectors in accordance with Fig. 3.1d. 11 and 1$\bar{1}$ vectors correspond to $\frac{1}{2}u \times [\bar{1}10]$ and $\frac{1}{2}u \times [001]$ direction vectors of the CsCl structure, respectively, u being the lattice constant. Since 01- and 11-type LEED spots match, within the accuracy of the measurement and similar to the case of AlCo domains (Section 3.2), with (32002)- and (42013)-type spots obtained from the pentagonal surface structure of the quasicrystal, u is ∼ 2.88 Å. Spot-profile analyses [44, 113] reveal domain diameters of ∼ 20 Å for a film thickness of 1.9 ML. At this stage of growth, spots still exhibit a circular shape with a constant diameter independent of the primary-electron energy.

For deposits of more than ∼ 4 ML, the surface structure is modified. This is apparent from the LEED patterns presented in Figs. 4.3b and c obtained from Ni deposits of 4.8 and 19 ML, respectively. An SEI pattern obtained from the surface after the deposition of 19 ML of Ni, revealing faint features, is shown in Fig. 4.3d. At polar angles in the range of $35 - 40°$, an increased intensity is found, giving rise

to a ring-like feature modulated roughly 10f. Between polar angles of $\sim$ 15 and
35° as well as for a polar angle of $\sim$ 45°, additional 10f rotational modulations are
observable.

SEMPA reveals that a 13.5 ML thick Ni overlayer exhibits out-of-plane magnetic
ordering. No in-plane signal is observed for deposits up to 13.5 and 19 ML using
SEMPA and longitudinal MOKE measurements, respectively.

Removing Ni by Ar$^+$-ion sputtering at room temperature and normal incidence
reproducibly leaves two different surface structures depending on the amount of
Ni previously deposited. If up to $\sim$ 2 ML of Ni have been deposited, the same

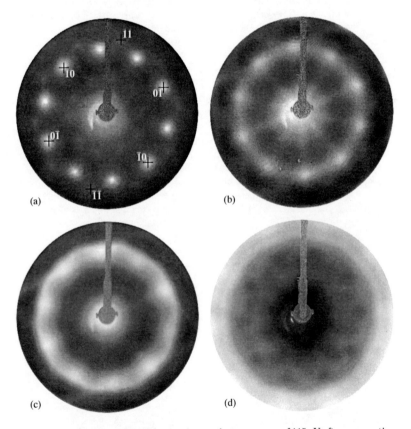

Fig. 4.3: LEED patterns obtained at a primary-electron energy of 115 eV after evaporating
(a) 1.9, (b) 4.8, and (c) 19 ML of Ni onto the pentagonal surface of i-Al-Pd-Mn. In
pattern (a), spots originating from one domain are marked with black crosses and indexed
according to Fig. 3.1d. The 11-type spots are second-order diffraction spots accompanying
the innermost spots (not marked). (d) SEI pattern recorded after evaporating 19 ML
of Ni.

surface structure is observed, consisting of five AlPd(113) domains rotated by 72°
with respect to each other, as found after sputtering the clean pentagonal surface
(cp. Figs. 4.4a and 4.1a). For pre-deposits of more than ∼ 2 ML, an SEI pattern
is found as displayed in Fig. 4.4b. It indicates the occurrence of five AlPd(110)
domains azimuthally rotated by 72° with respect to each other (cp. with Fig. 3.2b).
The 5f rotational symmetry of the pattern is a result of a tilt of the domains of
approximately $0.5 - 1°$ relative to the surface normal towards adjacent 3f-symmetry
axes of the i-substrate (Section 3.3). In both cases, AES measurements show comparable chemical compositions. Naturally, relative concentrations of Al, Mn, and
Pd which are also removed during completely removing Ni by means of sputtering
cannot be controlled. However, once either one or the other surface configuration
is established it was found to be stable against sputtering.

4.2 Discussion of the growth mode of Ni

As described in Section 3.3, five nominal AlPd(113) domains rotated by 72° with
respect to each other are detected after sputtering the clean surface. CsCl-type
AlPd domains are formed as a result of sputtering and their orientation results
from the optimum matching with the average structure of the i-substrate. During
annealing the sample at sufficiently high temperatures, the quasicrystalline surface
structure is restored due to the modified chemical composition caused by diffusion.
The occurrence of well-resolved Kikuchi bands [103, 105] originating from 2f- and
5f-symmetry planes of the i-structure (Fig. 4.1b) testifies for its high quality in a
near-surface region. The quality of the LEED pattern shown in Fig. 4.1c also points
to a well-established quasicrystalline order at the surface [98, 102].

The lack of decrease in intensities of the Mn $L_3M_{2,3}M_{4,5}$ Auger signal compared
to the bare substrate and the relatively weak intensities of the Ni signal for deposits

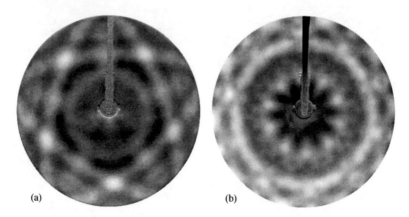

(a)   (b)

Fig. 4.4: SEI patterns obtained after completely removing Ni deposits of (a) 1.3 and (b)
3.2 ML by means of sputtering at normal incidence and room temperature.

of 0.3 ML indicate intermixing of these two elements (Fig. 4.2). Pd appears not to take part in this process which suggests that it is not present in the first atomic surface layer of the substrate as proposed by Gierer et al. [116] and confirmed by Bastasz [117]. The relatively low concentration of Mn in the sample suggests that Al also takes part in the intermixing process in order to explain the weak increase of the Ni signal at this stage of growth. For further deposition, Al migrates to growing Al-Ni layers which accounts for the weak growth of Ni signal intensities compared to the attenuation of Pd and Mn signal intensities.

Considering the mean free path of a few Å for the low-energy electrons [97, 99, 100], the observation of LEED spots originating from the quasicrystalline structure for deposits up to 1.5 ML of Ni suggests that intermixing and initial domain growth proceed rather homogeneously. Al-Ni domains have the same orientation and a similar lattice constant like the AlCo and Co domains (Section 3.2). Obviously, the arrangement of these CsCl-type domains is also determined by the optimum matching with the average structure of the i-substrate although the underlying surface region formed by intermixing is, since no surface structure is observed for submonolayer deposits of Ni, initially unordered. Shape and dependence on the primary-electron energy of spot profiles observed for a 1.9-ML deposit indicate a smooth surface.

Ni metal crystallises in the face-centred cubic structure with a lattice constant of 3.52 Å [129]. Ni-rich Al-Ni alloys possess the same crystal structure with a lattice constant of ~ 3.6 Å, while AlNi possesses the CsCl structure with a lattice constant of $2.8 - 2.9$ Å [115], in accordance with the present value. In addition, films of bcc Ni with lattice constants of $2.8 - 2.9$ Å have been reported (see, e.g., Refs. [130, 131]). Considering observations made with LEED and AES, CsCl-type Al-Ni domains, closely related to the AlNi alloy, form on the apparently unordered surface region due to Al diffusion from the substrate to the growing film. Hence, chemical and structural conditions are fulfilled for the growth of Ni in the bcc structure.

Compared to the interface between AlCo domains exposing their (110) faces and the pentagonal surface of i-Al-Pd-Mn, the interface between the Al-Ni domains, which also expose their (110) faces, and the substrate is less distinct (Fig. 4.5). On i-Al-Pd-Mn an Al-lean Al-Pd-Mn region is found, caused by Al diffusion to evaporated Ni. On this region the Al-Mn-Ni layer formed by intermixing, at which Pd did not take part, is present followed by the CsCl-type Al-Ni domains. Since the structure of the Al-Ni domains and the average substrate structure exhibit optimum matching, the Al-Mn-Ni layer in-between which transmits structural information most likely assumes the same structure and orientation like the domains. This is conceivable, since Mn-lean Al-Mn-Ni alloys have a CsCl-type structure and a lattice constant of $2.85 - 2.90$ Å [132]. Potentially, an Al-lean Al-Pd-Mn region may also assume the same structure and orientation like the Al-Ni domains. Definitely, however, the interface is formed similarly to the interface between the CsCl-type AlCo domains and the 5f-symmetry surface of i-Al-Pd-Mn (Section 3.4). The established structural CsCl(110) character of the interface acts as a precursor for the formation of AlPd(110) domains, if Ni is removed by means of sputtering, only for pre-deposits of more than ~ 2 ML indicating that the interface is, compared to smaller pre-deposits, more extensively developed at this stage of growth.

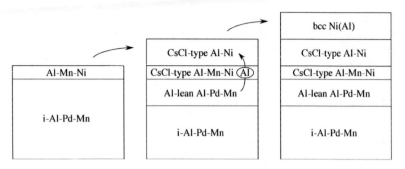

Fig. 4.5: Schematic illustration of the formation of the interface between evaporated Ni and i-Al-Pd-Mn. At the initial stage of growth, Ni intermixes with the substrate surface forming an unordered layer. For further Ni deposition, Al diffuses to growing Al-Ni domains which possess the CsCl structure and expose their (110) faces. As a result, an Al-lean Al-Pd-Mn near-surface region is formed. The initially unordered layer formed by intermixing assumes structure and orientation of the Al-Ni domains on which bcc Ni, potentially containing some Al, grows.

In comparison with LEED patterns obtained for a Ni deposit of 1.9 ML, patterns recorded for deposits of 4.8 and 19 ML show additional intensities pointing to a modified surface structure (Fig. 4.3a – c). Although less distinct, these intensities are similar to those found in patterns obtained from Co domains exhibiting a row structure at the surface (Section 3.5). Accordingly, the domains obtained for deposits of more than ~ 4 ML also exhibit a row structure at the surface. Similar to the case of Co, rows are parallel to $[\bar{1}11]$ and $[\bar{1}1\bar{1}]$ directions.

As visible in the SEI pattern displayed in Fig. 4.3d, no distinct features which can usually be assigned to features of a layer consisting of five CsCl-type domains exposing their (110) faces and rotated by 72° with respect to each other have been formed for deposits of 19 ML (cp. with Fig. 4.4b). However, a 10f rotational symmetry is discernible which originates from the combination of a 5f (pentagonal surface) and a 2f [cubic (110) face] rotational symmetry. Increased intensities in the range of $35 - 40°$ can be attributed to [111] and $[11\bar{1}]$ directions of the cubic structure, while increased intensities attributed to directions with higher indices (e.g., $\langle 11i \rangle$ directions, $i = 2, 3$, and 4) usually found in Kikuchi bands originating from {110} planes are missing. Increased intensities observed at $\sim 45°$ can be assigned to [100] and [010] directions.

For both Ni films on Fe(100) [130] and GaAs(100) [131], it has been observed that the growth of bcc Ni is followed by the growth of a novel structure. In the case of an Fe(100) substrate, bcc Ni is observed for films thinner than ~ 6 layers. For higher coverages up to ~ 20 layers, a $c(2 \times 2)$ structure forms. For even larger Ni deposits up to ~ 100 layers, Wang et al. have reported a "complicated", undetermined structure [130]. In the case of a GaAs(100) substrate, Tian et al. have mentioned that a "more complicated" and also undetermined structure develops on an ~ 3.5 nm thick bcc Ni film [131]. Ni domains formed on the pentagonal surface of i-Al-Pd-Mn for deposits of more than ~ 4 ML possess, as observable with LEED, a surface structure consisting of rows which impedes the observation of distinct features related to CsCl-type (110) domains by means of SEI.

Different to Ni films deposited on GaAs(100) [131] which show in-plane magnetic ordering, out-of-plane magnetic ordering is observed for an $\sim$ 13.5 ML thick Ni deposit. The existence of magnetic ordering at this stage of growth indicates that the surface region mainly consists of Ni, since only Ni-rich Al-Ni alloys show magnetic ordering at room temperature [133]. Therefore, considering the corresponding surface texture observed with LEED, one can conclude that Ni domains, potentially containing some Al, with a bcc structure grow on Al-Ni domains. Obviously, the CsCl structure of the Al-Ni domains stabilises the extraordinary bcc phase of Ni.

Chapter 5

Fe deposition on the pentagonal surface of icosahedral Al-Pd-Mn

5.1   Experimental results

Fig. 5.1a shows a LEED pattern obtained from the 5f-symmetry surface of the sputter-annealed i-Al-Pd-Mn quasicrystal at a primary-electron energy of 74 eV. LEED spots are numbered according to Figs. 3.1a and 4.1c and specified in Table 3.1. In Fig. 5.1b, the corresponding SEI pattern displaying the 5f symmetry of the surface is presented.

The intensities of LEED spots decrease markedly and spots show some broadening as soon as minute amounts of Fe [$\sim$ 0.1 monolayer equivalents (MLE)[1]] are deposited onto the quasicrystalline surface kept at $\sim$ 340 K.[2] After the evaporation of 2 MLE, no spots originating from the substrate are observable anymore. In addition to this obliteration of the structure, AES measurements show a significantly smaller increase of the intensities of the Fe LMM signals during the deposition onto the pentagonal surface of i-Al-Pd-Mn compared to the growth of Fe on polycrystalline Cu (Fig. 5.2). Furthermore, the intensities of the Pd and Mn Auger signals do not decrease and the Al LMM signal only shows a relatively small decrease during the initial Fe evaporation ($\lesssim$ 2 MLE). For deposits larger than some MLE, the Al signal intensities cannot be reliably analysed due to the overlap of the Al LMM and Fe MNN Auger signals.

New LEED spots originating from a different surface structure appear for the deposition of $\sim$ 4 MLE of Fe. A corresponding LEED pattern recorded after the deposition of 8 MLE at a primary-electron energy of 162 eV is shown in Fig. 5.3a. Spots are positioned on two circles, while neighbouring spots on each circle are azimuthally separated by 36°. By varying the primary-electron energy, several other diffraction spots become visible which reveal an arrangement of five domains each possessing a surface structure derived from a CsCl-type (110) surface and rotated by 72° increments. Spots belonging to one domain are marked with black

---

[1]1 MLE $\approx$ 2.04 Å, considering structure and orientation of the growing film. CsCl-type Al-Fe and bcc Fe domains have a lattice constant of $\sim$ 2.88 Å and expose their (110) faces.

[2]Investigations revealed a similar growth mode for the sample kept at room temperature during evaporation.

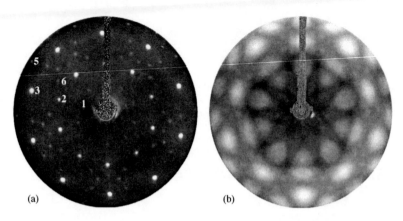

Fig. 5.1: (a) LEED pattern obtained from the 5f-symmetry surface of i-Al-Pd-Mn at a primary-electron energy of 74 eV. LEED spots are numbered on the right-hand side in accordance with Figs. 3.1a and 4.1c. (11001)-type spots corresponding to number 1 are hardly discernible and (42002)-type spots corresponding to number 4 are not located in the range of the display screen. The observability of less spots compared to Figs. 3.1a and 4.1c is mainly a result of the lower primary-electron energy and the shorter exposure time. (b) SEI pattern recorded from the same surface.

crosses and labelled with the components of corresponding reciprocal surface-lattice vectors in accordance with Figs. 3.1d and 4.3a. Accordingly, 11 and $1\bar{1}$ vectors correspond to $\frac{1}{2}v \times [\bar{1}10]$ and $\frac{1}{2}v \times [001]$ direction vectors of the CsCl structure, respectively, v being the lattice constant. 01- and 11-type LEED spots match, within the accuracy of the measurement and similar to AlCo (Section 3.2) and Al-Ni (Section 4.1) domains, with (32002)- and (42013)-type spots obtained from the pentagonal surface of i-Al-Pd-Mn. Consequently, v is ~ 2.88 Å. The angular widths of LEED spots indicate [44, 113] domain diameters of ~ 35 Å for 10-MLE and ~ 15 Å for 35-MLE thick films.

For Fe coverages higher than ~ 4 MLE, longitudinal MOKE measurements bring up evidence for in-plane magnetic ordering in the growing film. Fig. 5.4 presents a diagram of the saturation magnetisation as a function of the Fe deposit. In the inset, MOKE signals as a function of the applied magnetic field obtained from Fe deposits of about 3, 6, 11, and 17 MLE are plotted. The magnetic field was applied with respect to the sample orientation identifiable in Fig. 5.3 along the horizontal. Consequently, the [001] directions in the CsCl(110)-type domains span angles of about 16.7, 52.7, 88.7, 124.7, and 160.7° with the applied field. The linear fit of the saturation magnetisation measured for films thicker than 6 MLE suggests a magnetic dead layer at the interface of ~ 2.5 MLE.

A small displacement of the LEED spots in polar and azimuthal directions is observed for Fe coverages higher than ~ 8 MLE. A corresponding pattern is shown in Fig. 5.3b. Similar to the pattern presented in Fig. 5.3a, the inner spots are positioned on a circle. However, neighbouring spots on this circle are azimuthally

Fig. 5.2: Fe concentrations in a near-surface region as a function of Fe deposited onto the pentagonal surface of i-Al-Pd-Mn, Al(100) [134], and polycrystalline Cu obtained from the LMM Auger signal and considering the appropriate relative Auger sensitivities of the elements [98]. The values of the Fe deposit correspond to a film thickness determined by assuming a layer-by-layer growth mode.

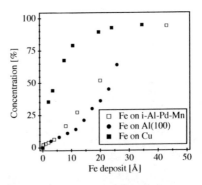

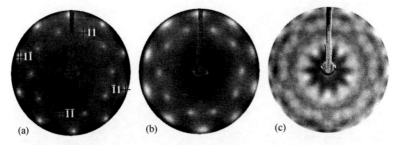

Fig. 5.3: LEED patterns obtained at a primary-electron energy of 162 eV after evaporating (a) 8 and (b) 32 MLE of Fe onto the clean pentagonal surface of i-Al-Pd-Mn. (c) SEI pattern obtained after the deposition of 32 MLE of Fe.

separated by alternating angles of approximately 33 and 39°. Furthermore, the outer spots appear on two circles with slightly different radii with every second spot being on the same circle. Spots on each circle are found in 72° increments. These new positions of the LEED spots can be accounted for by a displacement of the five domains by a polar inclination of ∼ 0.5° away from the surface normal towards adjacent 3f-symmetry axes.

Fig. 5.3c shows an SEI pattern obtained after evaporating 32 MLE of Fe on the quasicrystalline surface. This pattern is distinctly different than that observed for the clean quasicrystal and shows features that originate from the five CsCl-type domains exposing their (110) faces (cp. with Figs. 3.2b and 4.4b). The 5f rotational symmetry is a result of the polar inclination of the domains.

Fig. 5.5 presents patterns recorded after removing deposits of 2 and 8 MLE of Fe by means of sputtering revealing the presence of five CsCl-type AlPd(113) and AlPd(110) domains, respectively, azimuthally rotated by 72° with respect to each other (cp. with Figs. 3.2 and 4.4). A domain structure consisting of domains exposing their (113) faces is obtained as a result of removing Fe deposits of a thickness up to ∼ 4 MLE by means of sputtering. For larger pre-deposits, domains exposing their (110) faces are observed which are tilted by approximately $0.5 - 1°$

Fig. 5.4: Dependence of saturation mag-netisation and Fe deposit. The line is a fit of the data points recorded for Fe deposits of more than 6 MLE. The inset shows MOKE hysteresis loops obtained from Fe deposits of about 3, 6, 11, and 17 MLE. The axis corresponding to the magnetisa-tion (applied magnetic field) runs in verti-cal (horizontal) direction and the point of origin is marked with a cross. The coercive field is about 100 Oe ($\approx$ 7958 A/m) for all measured Fe deposits.

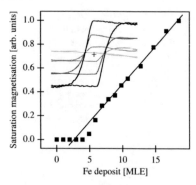
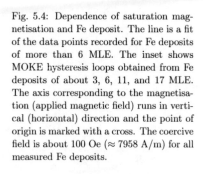

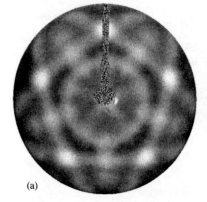
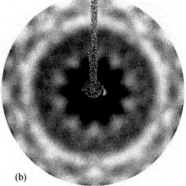

(a)

(b)

Fig. 5.5: SEI patterns obtained after completely removing Fe. Pre-deposits were (a) 2 and (b) 8 MLE.

relative to the surface normal towards adjacent 3f-symmetry axes (see Section 3.3). This tilt gives rise to an SEI pattern which exhibits 5f rather than 10f rotational symmetry (cp. with Figs. 3.2b, 4.4b, and 5.3c). During completely removing Fe, the relative concentrations of Al, Mn, and Pd also removed cannot be controlled. However, both observed configurations were found to be stable against sputtering at room temperature and normal incidence, as performed in this investigation.

5.2   Discussion of the growth mode of Fe

The quality of the LEED and SEI patterns shown in Fig. 5.1 point to a well-ordered quasicrystalline surface structure [98, 102]. Fe intermixes with this surface at the initial stage of growth as indicated by the non-decreasing AES signal intensities of Pd and Mn, the small decrease of the Al signal, as well as the small increase of the Fe

Auger signals compared to the case of the rather uniform covering of polycrystalline Cu with evaporated Fe and similar to the growth of Fe on Al(100) [134] (see Fig. 5.2). The observed broadening of LEED spots originating from the i-substrate during the initial Fe deposition suggests a gradual disappearance of quasicrystalline areas, i.e., inhomogeneous intermixing. For further deposition, Al migrates to growing Al-Fe layers which accounts for the relatively weak growth of the Fe signal intensities. The magnetic dead layer of ~ 2.5 MLE is a result of intermixing and Al diffusion.

Al-Fe domains formed for Fe deposits between 4 and 8 MLE exhibit the same orientation and a similar lattice constant like AlCo, Co (Section 3.2), and Al-Ni (Section 4.1) domains. Similar to the growth of Al-Ni domains, the optimum matching of the average structure of the i-substrate and the CsCl-type domains determines the domain orientation in spite of the presence of an initially unordered layer formed by intermixing.

Fe metal crystallises, at room temperature, in the bcc structure with a lattice constant of 2.87 Å [129]. Fe-rich Al-Fe alloys as well as AlFe possess CsCl-type structures with lattice constants of ~ 2.9 Å [115, 133]. Consequently, it is conceivable that bcc Fe domains grow on CsCl-type Al-Fe domains which grow on the seemingly unordered layer formed by intermixing. Atomic layers formed for deposits of more than ~ 2.5 MLE appear to be Fe-rich, since only CsCl-type Fe_3Al and bcc Fe, potentially containing some Al, are ferromagnetic [133].

Al-Fe domains are tilted, both in direction and extent, in correspondence to AlPd(110) domains found after removing pre-deposits of more than ~ 4 MLE of Fe, ~ 1 ML of Co (Section 3.3), and ~ 2 ML of Ni (Section 4.1) only for layers thicker than ~ 8 MLE. Correspondingly, this tilt is most likely induced by the domain-domain interfaces although structural strain between Fe-rich Al-Fe alloy structures may also contribute. The growth mode is three-dimensional at this stage of growth since domain dimensions decrease as a function of increasing Fe deposit.

The interface between Al-Fe domains and the 5f-symmetry surface of i-Al-Pd-Mn is formed similarly to the interface between Al-Ni domains and the same substrate (Section 4.2). An Al-lean Al-Pd-Mn region is formed as a result of Al diffusion from the quasicrystal to Al-Fe domains growing on the Al-Fe-Mn-Pd layer formed by intermixing. Since structural information between the Al-Fe domains and i-Al-Pd-Mn has to be transmitted in order to allow optimum matching, the initially unordered Al-Fe-Mn-Pd layer most likely assumes the same structure and orientation like the Al-Fe domains. Al-lean Al-Pd-Mn may also assume structure and orientation of the domains. Definitely, however, the interface between the i- and the cubic structure is formed similarly to the interface between AlCo and the pentagonal surface of i-Al-Pd-Mn. Consequently, the structural CsCl(110) character of the interface established for Fe deposits of more than ~ 4 MLE also acts as a precursor for the formation of AlPd(110) domains during the removal of Fe by means of sputtering.

Chapter 6

Comparison of the growth modes, conclusion, and outlook

6.1 Comparison of the growth modes of Co, Ni, and Fe

In the two cases of Ni and Fe deposition on the pentagonal surface of i-Al-Pd-Mn, the interfaces are formed by intermixing and Al diffusion. At the initial stage of growth, Ni as well as Fe intermix with the substrate surface forming unordered surface layers. Ni intermixes with Al and Mn, while Fe also intermixes with Pd suggesting a stronger intermixing of Fe with the substrate compared to Ni. For further deposition, Al migrates from the substrate to Al-Ni and Al-Fe layers which grow on the layers formed by intermixing. Al-Ni and Al-Fe domains, which possess a CsCl-type structure and expose their (110) faces, are formed for Ni deposits of ~ 1 ML and Fe deposits of ~ 4 MLE, respectively. The thicker layer formed by intermixing may impede the formation of ordered Al-Fe domains for smaller deposits. The orientation of both CsCl-type Al-Ni and Al-Fe domains is determined by the optimum matching with the periodic, discrete average substrate structure which suggests that the initially unordered layers formed by intermixing assume the same structure like the domains. For further evaporation, both Ni and Fe, potentially containing some Al, grow on the CsCl-type domains in the bcc structure.

AlNi, AlFe, and AlCo possess heats of formation of about -0.7, -0.3, and -0.6 eV/atom, respectively [135]. This suggests that the formation of the interface between Co and the substrate is governed, similarly to the cases of Ni and Fe, by intermixing and diffusion. In contrast, evaporated Co does not intermix with the substrate normal to the surface. An AlCo atomic layer, which consists of domains possessing a structure derived from a CsCl-type (110) surface, is formed for submonolayer deposits. On the AlCo domains, Co grows in the bcc structure. The AlCo layer appears to be a barrier for Al diffusion. The orientation of domains with respect to the substrate, which is similar to the orientation of Al-Ni and Al-Fe domains, is also determined by the optimum matching with the average structure of the substrate.

Compared to Co, larger deposits are needed in the cases of Ni and Fe in order to obtain ordered domains on which Ni and Fe grow in the bcc structure, respectively.

Obviously, it is the tendency of the adsorbates to alloy with the substrate which determines the amount of deposited atoms at which a domain structure is formed at the surface. This tendency is highest for Co, followed by Ni and Fe.

Co and Al-Ni domains have diameters of ~ 20 Å for deposits of ~ 2 ML, while Fe domains still have a diameter of ~ 35 Å for a 10-MLE deposit. Diameters of Fe domains decrease with increasing deposits which indicates three-dimensional growth. In addition, Fe domains show a small tilt for deposits larger than ~ 8 MLE which is most likely induced by domain boundaries, while no tilt is observed in the cases of Co and Ni domains. In contrast to Fe, the surface structure of Co and Ni domains is modified for deposits of more than ~ 3 and 4 ML, respectively. Atomic row structures are formed in both cases.

Removing, by means of sputtering, Co, Ni, and Fe deposits of at least 1 ML, 2 ML, and 4 MLE, respectively, reveals a structure consisting of five CsCl-type AlPd(110) domains instead of a structure consisting of AlPd(113) domains observed after sputtering the clean substrate surface or removing smaller deposits. The structural CsCl(110) character of the interface, which is, with respect to increasing deposits, first established for Co followed by Ni and Fe, favours the formation of AlPd(110) domains during sputtering.

Ni and possibly Al-Ni domains exhibit out-of-plane magnetic ordering, while in-plane magnetic ordering is observed in Co and Al-Co as well as Fe and Al-Fe domains. For sufficiently high Co coverages, the entire film is ferromagnetic. In contrast, a magnetic dead layer of ~ 2.5 MLE is found at the interface between Fe and the quasicrystal due to intermixing.

The coincidence reciprocal-lattice-planes model of Widjaja and Marks successfully describes the alignment of CsCl-type domains on i-Al-Cu-Fe [83,86,136], which is isostructural to i-Al-Pd-Mn [137]. However, this model assumes a discontinuous quasicrystal-crystal interface, while the approach to explain the alignment of CsCl-type domains on d-Al-Co-Cu, d-Al-Co-Ni, and i-Al-Pd-Mn with the optimum matching of the average structures [85] would allow to describe the interface by a continuous transition from the quasicrystalline to the crystalline structure. This would be consistent with the observation that long-range order is lost in an initially quasicrystalline surface layer for submonolayer deposits of Co. Due to the similarity in symmetry and absolute measures of the surfaces of i- and d-quasicrystals [85], one may expect comparable growth modes like those of Co, Ni, and Fe on the pentagonal surface of i-Al-Pd-Mn for the deposition of the same elements on the quasiperiodic surfaces of quasicrystals like i-Al-Cu-Fe, i-Al-Cu-Ru, i-Al-Pd-Re, d-Al-Co-Ni, and d-Al-Co-Cu.

6.2   Conclusion

The growth of Co, Ni, and Fe upon the deposition on the pentagonal surface of an i-Al-Pd-Mn quasicrystal has been investigated by means of LEED, SEI, AES, SEMPA, and MOKE measurements.

The deposition of 0.34 ML of Co results in the loss of long-range order in an initially quasicrystalline surface region accounted for by the already incipient formation of the interface between Co and the quasicrystal. No intermixing of Co with the substrate normal to the surface is observed. For a deposit of 0.67 ML, an atomic layer consisting of five AlCo domains with nm dimensions is formed. The domains possess a structure derived from a CsCl-type (110) surface and are rotated by 72° with respect to each other. The orientational relationship between the domains and the substrate results from the optimum matching of the average structures, which would, however, allow four different alignments with respect to the surface. The observed alignment is favoured due to alignment-dependent surface and domain-domain interface energies. For further deposition, Co in the bcc structure epitaxially grows on the AlCo domains up to a film thickness of ~ 3 ML. For larger deposits, a modified surface structure is formed on the Co domains which consists of an arrangement of one-dimensional atomic rows parallel to $[\bar{1}11]$ and $[\bar{1}1\bar{1}]$ directions. The entire Co film exhibits in-plane magnetic ordering for deposits of more than ~ 1.5 ML. The growth mode of Co may be referred to as quasi-epitaxial, since, due to the close structural relationship of the substrate and the domains, only minor atomic displacements are expected for the interface formation.

At the initial stage of growth, Ni intermixes with the substrate surface forming an unordered Al-Mn-Ni layer. For further deposition, Al migrates from the substrate to five Al-Ni domains which grow on this layer, possess a surface structure derived from a CsCl-type (110) surface, have nm dimensions, and are rotated by 72° with respect to each other. The orientational relationship between the Al-Ni domains and the substrate is a result of the optimum matching of the average structures. The Al-Mn-Ni layer, which transmits structural information, most likely assumes the same structure and orientation like the Al-Ni domains on which Ni in the bcc structure, potentially containing some Al, is growing for further deposition. For deposits of more than ~ 4 ML, Ni domains exhibit a row structure at the surface. Ni films exhibit out-of-plane magnetic ordering.

The interface between Fe and the substrate is formed similarly to the case of Ni. Al migrates from the quasicrystal to an Al-Fe layer growing on an Al-Fe-Mn-Pd layer formed by intermixing at the initial stage of growth. For sufficiently high Fe coverages of ~ 4 MLE, the Al-Fe layer consists of five CsCl-type domains which expose their (110) faces, are rotated by 72° with respect to each other, and have nm dimensions. The orientation of these domains is determined by the optimum matching with the average structure of the i-substrate. The layer formed by intermixing appears to assume the same structure and orientation like the domains on which bcc Fe grows for further deposition. For deposits of more than ~ 8 MLE, Fe domains exhibit a small tilt induced by domain boundaries. In-plane magnetic ordering is found for coverages thicker than 4 MLE. A magnetic dead layer of ~ 2.5 MLE is formed at the interface due to intermixing.

Although some differences between the growth modes of Co, Ni, and Fe on the pentagonal surface of i-Al-Pd-Mn exist, the formation of the interfaces is similar. Ordered domains are formed due to alloying of the adsorbates with Al and the orientation of the CsCl-type domains results from the optimum matching with the periodic, discrete average structure of the substrate.

Ion sputtering i-Al-Pd-Mn results in a modified chemical composition in a near-surface region giving rise to the formation of CsCl-type AlPd. Provided that sputtering is isotropic, this AlPd layer consists of $5n$ domains, whereas $n \in \mathbf{N}$ and domains within one set are azimuthally rotated by 72° with respect to each other, conserving the substrate symmetry on a global scale. The fundamental orientation of these domains with respect to the substrate is governed by the optimum matching of the average structures allowing four different alignments ($n \in \{1, 2, 3, 4\}$). Not all of these alignments are found and the fundamental orientation of each existing alignment may slightly be modified due to an alignment-selecting sputtering mechanism and different surface as well as domain-domain interface alignments. Sputtering the clean substrate surface and removing, by means of sputtering, Co, Ni, and Fe deposits of less than ~ 1 ML, 2 ML, and 4 MLE, respectively, reveals a structure consisting of nominal AlPd(113) domains, possibly corresponding to two similar alignments. If larger deposits are removed, the structural CsCl(110) character of the interface favours the formation of AlPd(110) domains.

6.3   Outlook

Co, Ni, and Fe evaporated onto the 5f-symmetry surface of i-Al-Pd-Mn form ferromagnetic films consisting of crystalline domains with nm dimensions and well-defined orientations mediated by the substrate symmetry. Future investigations may reveal how the domains are arranged with respect to each other, in particular to whether they form a quasiperiodic superlattice on the quasicrystalline substrate. Furthermore, it could be investigated to what extent individual domains are discrete regarding atomic structure, electronic structure, and magnetism, how these properties may be controlled by modifying the growth mode of domains by varying deposition rate and sample temperature during deposition, and how the arrangement of domains influences electronic and magnetic properties. The deposition of atoms on particular surfaces of approximants may provide an approach for generating a periodic superlattice of crystalline domains with nm dimensions and well-defined orientations, mediated by the local atomic structure corresponding to that of a quasicrystalline structure.

The optimum matching of the CsCl-type structure of domains and the discrete average structure of i-Al-Pd-Mn indicates the close structural relationship of the two structures. To shed light on this relationship and describing the interface formation, a detailed characterisation of the discrete average structure of i-Al-Pd-Mn and model calculations, which require a detailed description of the atomic structure in a near-surface region, would be desirable. They may disclose details of the nature of the quasicrystalline structure.

Bibliography

[1] D. Hume, A treatise of human nature, Book I, Part IV, Section VII, http://www.gutenberg.org/etext/4705.

[2] D. Hume, in: T. Lipps (Ed.), David Hume's Traktat über die menschliche Natur — Ein Versuch die Methode der Erfahrung in die Geisteswissenschaft einzuführen — Über den Verstand, Verlag von Leopold Voss, Leipzig, Hamburg, 1912, p. 346.

[3] D. Shechtman, I. Blech, D. Gratias, J.W. Cahn, Phys. Rev. Lett. 53 (1984) 1951.

[4] D. Levine, P.J. Steinhardt, Phys. Rev. B 34 (1986) 596.

[5] G. Chapuis, Cryst. Eng. 6 (2003) 187.

[6] K.H. Wiederkehr, Centaurus 21 (1977) 278.

[7] V. Goldschmidt, C. Palache, M. Peacock, Neues Jahrbuch für Mineralogie, Geologie und Paläontologie A 63 (1931) 1.

[8] E. Brouns, J.W. Visser, P.M. de Wolff, Acta Cryst. 17 (1964) 614.

[9] W.J. Schutte, J.L. de Boer, Acta Cryst. B 44 (1988) 486.

[10] M. Dusek, G. Chapuis, M. Meyer, V. Petricek, Acta Cryst. B 59 (2003) 337.

[11] T. Janssen, A. Janner, Adv. Phys. 36 (1987) 519.

[12] M.I. McMahon, L.F. Lundegaard, C. Hejny, S. Falconi, R.J. Nelmes, Phys. Rev. B 73 (2006) 134102.

[13] P.M. de Wolff, T. Janssen, A. Janner, Acta Cryst. A 37 (1981) 625.

[14] G. Venkataraman, D. Sahoo, V. Balakrishnan, Beyond the crystalline state — An emerging perspective, Springer, Berlin, Heidelberg, 1989, p. 7.

[15] W. Steurer, T. Haibach, in: Z.M. Stadnik (Ed.), Physical properties of quasicrystals, Springer, Berlin, Heidelberg, 1999, p. 51.

[16] M. de Boissieu, S. Francoual, Z. Kristallogr. 220 (2005) 1043.

[17] S. Francoual, F. Livet, M. de Boissieu, F. Yakhou, F. Bley, A. Létoublon, R. Caudron, J. Gastaldi, Phys. Rev. Lett. 91 (2003) 225501.

[18] W. Steurer, T. Haibach, Acta Cryst. A 55 (1999) 48.

[19] A.I. Goldman, R.F. Kelton, Rev. Mod. Phys. 65 (1993) 213.

[20] A.P. Tsai, in: Z.M. Stadnik (Ed.), Physical properties of quasicrystals, Springer, Berlin, Heidelberg, 1999, p. 5.

[21] M. Feuerbacher, C. Thomas, K. Urban, in: H.-R. Trebin (Ed.), Quasicrystals — Structure and physical properties, Wiley, Weinheim, 2003, p. 2.

[22] T. Gödecke, R. Lück, Z. Metallkd. 86 (1995) 109.

[23] G. Busch, H. Schade, Vorlesungen über Festkörperphysik, Birkhäuser, Basel, 1973, p. 59; R.A. Swalin, Thermodynamics of solids, Wiley, New York, 1972, p. 45.

[24] M. de Boissieu, Phil. Mag. 86 (2006) 1115.

[25] P.A. Bancel, P.A. Heiney, Phys. Rev. B 33 (1986) 7917.

[26] W. Hume-Rothery, J. Inst. Met. 35 (1926) 295.

[27] H. Jones, Proc. Phys. Soc. London 49 (1937) 250.

[28] Z.M. Stadnik, in: Z.M. Stadnik (Ed.), Physical properties of quasicrystals, Springer, Berlin, Heidelberg, 1999, p. 257.

[29] A. Suchodolskis, W. Assmus, B. Čechavičius, J. Dalmas, L. Giovanelli, M. Göthelid, U.O. Karlsson, V. Karpus, G. Le Lay, R. Sterzel, E. Uhrig, Appl. Surf. Sci. 212 − 213 (2003) 485.

[30] E. Rotenberg, W. Theis, K. Horn, Prog. Surf. Sci. 75 (2004) 237.

[31] M. Gardner, Sci. Am. 236 (1977) 110.

[32] C. Sire, in: F. Hippert, D. Gratias (Eds.), Lectures on quasicrystals, Les Editions de Physique, Les Ulis, 1994, p. 505.

[33] C.L. Henley, in: D.P. DiVincenzo, P.J. Steinhardt (Eds.), Quasicrystals — The state of the art, World Scientific, Singapore, 1999, p. 459.

[34] Ö. Rapp, in: Z.M. Stadnik (Ed.), Physical properties of quasicrystals, Springer, Berlin, Heidelberg, 1999, p. 127.

[35] M.A. Chernikov, A. Bianchi, H.R. Ott, Phys. Rev. B 51 (1995) 153; K. Giannò, A.V. Sologubenko, M.A. Chernikov, H.R. Ott, I.R. Fisher, P.C. Canfield, Mater. Sci. Eng. 294 − 296 (2000) 715.

[36] J.-M. Dubois, P. Brunet, W. Costin, A. Merstallinger, J. Non-Cryst. Solids 334 & 335 (2004) 475.

[37] V. Fournée, A.R. Ross, T.A. Lograsso, J.W. Evans, P.A. Thiel, Surf. Sci. 537 (2003) 5.

[38] N. Rivier, in: A.I. Goldman, D.J. Sordelet, P.A. Thiel, J.-M. Dubois (Eds.), New horizons in quasicrystals — Research and applications, World Scientific, Singapore, 1997, p. 188.

[39] Ph. Ebert, M. Yurechko, F. Kluge, K. Horn, K. Urban, in: H.-R. Trebin (Ed.), Quasicrystals — Structure and physical properties, Wiley, Weinheim, 2003, p. 572.

[40] R. Kelly, Nucl. Instrum. Methods Phys. Res. B 39 (1989) 43.

[41] A.R. Kortan, R.S. Becker, F.A. Thiel, H.S. Chen, Phys. Rev. Lett. 64 (1990) 200.

[42] Th.M. Schaub, D.E. Bürgler, H.-J. Güntherodt, J.B. Suck, M. Audier, Appl. Phys. A 61 (1995) 491.

[43] M. Zurkirch, B. Bolliger, M. Erbudak, A.R. Kortan, Phys. Rev. B 58 (1998) 14113; M. Shimoda, J.Q. Guo, T.J. Sato, A.-P. Tsai, Surf. Sci. 454 − 456 (2000) 11; C. Cecco, M. Albrecht, H. Wider, A. Maier, G. Schatz, G. Krausch, P. Gille, J. Alloys Comp. 342 (2002) 437; T. Flückiger, T. Michlmayr, C. Biely, R. Lüscher, M. Erbudak, Appl. Surf. Sci. 212 − 213 (2003) 43; T. Flückiger, Crystal-quasicrystal formations on Al-Co-Ni, http://e-collection.ethbib.ethz.ch/show?type=diss&nr=15308.

[44] M. Erbudak, J.-N. Longchamp, Y. Weisskopf, Turk. J. Phys. 29 (2005) 277.

[45] D. Naumović, P. Aebi, L. Schlapbach, C. Beeli, Mater. Sci. Eng. A 294 − 296 (2000) 882.

[46] B. Bolliger, M. Erbudak, A. Hensch, D.D. Vvedensky, Mater. Sci. Eng. A 294 − 296 (2000) 859.

[47] B. Bolliger, M. Erbudak, D.D. Vvedensky, M. Zurkirch, A.R. Kortan, Phys. Rev. Lett. 80 (1998) 5369.

[48] Z. Shen, M.J. Kramer, C.J. Jenks, A.I. Goldman, T. Lograsso, D. Delaney, M. Heinzig, W. Raberg, P.A. Thiel, Phys. Rev. B 58 (1998) 9961.

[49] T.C.Q. Noakes, P. Bailey, C.F. McConville, C.R. Parkinson, M. Draxler, J. Smerdon, J. Ledieu, R. McGrath, A.R. Ross, T.A. Lograsso, Surf. Sci. 583 (2005) 139.

[50] M. Shimoda, T.J. Sato, A.-P. Tsai, J.Q. Guo, J. Alloys Comp. 342 (2002) 441.

[51] V. Fournée, T.C. Cai, A.R. Ross, T.A. Lograsso, J.W. Evans, P.A. Thiel, Phys. Rev. B 67 (2003) 33406.

[52] V. Fournée, H.R. Sharma, M. Shimoda, A.P. Tsai, B. Unal, A.R. Ross, T.A. Lograsso, P.A. Thiel, Phys. Rev. Lett. 95 (2005) 155504.

[53] T. Flückiger, Y. Weisskopf, M. Erbudak, R. Lüscher, A.R. Kortan, Nano Lett.
     3 (2003) 1717.

[54] B. Bolliger, V.E. Dmitrienko, M. Erbudak, R. Lüscher, H.-U. Nissen,
     A.R. Kortan, Phys. Rev. B 63 (2001) 52203.

[55] R. Lüscher, M. Erbudak, Y. Weisskopf, Surf. Sci. 569 (2004) 163.

[56] R. Lüscher, T. Flückiger, M. Erbudak, Ferroelectrics 305 (2004) 245.

[57] T. Cai, J. Ledieu, R. McGrath, V. Fournée, T. Lograsso, A. Ross, P. Thiel,
     Surf. Sci. 526 (2003) 115.

[58] M. Shimoda, T.J. Sato, A.P. Tsai, J.Q. Guo, Phys. Rev. B 62 (2000) 11288;
     M. Shimoda, J.Q. Guo, T.J. Sato, A.P. Tsai, Surf. Sci. 482 − 485 (2001) 784.

[59] M. Shimoda, J.Q. Guo, T.J. Sato, A.P. Tsai, Jpn. J. Appl. Phys. 40 (2001)
     6073.

[60] K.J. Franke, H.R. Sharma, W. Theis, P. Gille, Ph. Ebert, K.H. Rieder, Phys.
     Rev. Lett. 89 (2002) 156104.

[61] E.J. Cox, J. Ledieu, V.R. Dhanak, S.D. Barrett, C.J. Jenks, I. Fisher,
     R. McGrath, Surf. Sci. 566 − 568 (2004) 1200.

[62] J. Ledieu, C.A. Muryn, G. Thornton, R.D. Diehl, T.A. Lograsso,
     D.W. Delaney, R. McGrath, Surf. Sci. 472 (2001) 89.

[63] R. McGrath, J. Ledieu, E.J. Cox, S. Haq, R.D. Diehl, C.J. Jenks, I. Fisher,
     A.R. Ross, T.A. Lograsso, J. Alloys Comp. 342 (2002) 432.

[64] J.T. Hoeft, J. Ledieu, S. Haq, T.A. Lograsso, A.R. Ross, R. McGrath, Phil.
     Mag. 86 (2006) 869.

[65] J.A. Smerdon, J. Ledieu, J.T. Hoeft, D.E. Reid, L.H. Wearing, R.D. Diehl,
     T.A. Lograsso, A.R. Ross, R. McGrath, Phil. Mag. 86 (2006) 841.

[66] Y. Weisskopf, S. Burkardt, M. Erbudak, J.-N. Longchamp, Surf. Sci., in press.

[67] J. Ledieu, J.T. Hoeft, D.E. Reid, J.A. Smerdon, R.D. Diehl, T.A. Lograsso,
     A.R. Ross, R. McGrath, Phys. Rev. Lett. 92 (2004) 135507; J. Ledieu,
     J.T. Hoeft, D.E. Reid, J.A. Smerdon, R.D. Diehl, N. Ferralis, T.A. Lograsso,
     A.R. Ross, R. McGrath, Phys. Rev. B 72 (2005) 35420.

[68] M. Bielmann, A. Barranco, P. Ruffieux, O. Gröning, R. Fasel, R. Widmer,
     P. Gröning, Adv. Eng. Mater. 7 (2005) 392.

[69] D. Reid, J.A. Smerdon, J. Ledieu, R. McGrath, Surf. Sci., in press.

[70] H.R. Sharma, M. Shimoda, V. Fournée, A.R. Ross, T.A. Lograsso, A.P. Tsai,
     Appl. Surf. Sci. 241 (2005) 256.

[71] R. Bastasz, J.A. Whaley, T.A. Lograsso, C.J. Jenks, Phil. Mag. 86 (2006) 855.

[72] Y. Weisskopf, R. Lüscher, M. Erbudak, Surf. Sci. 578 (2005) 35.

[73] Y. Weisskopf, M. Erbudak, J.-N. Longchamp, T. Michlmayr, Surf. Sci. 600 (2006) 2594.

[74] S.-L. Chang, W.B. Chin, C.-M. Zhang, C.J. Jenks, P.A. Thiel, Surf. Sci. 337 (1995) 135.

[75] M. Shimoda, T.J. Sato, A.-P. Tsai, J.Q. Guo, Surf. Sci. 507 − 510 (2002) 276.

[76] J. Ledieu, V.R. Dhanak, R.D. Diehl, T.A. Lograsso, D.W. Delaney, R. McGrath, Surf. Sci. 512 (2002) 77.

[77] L. Leung, J. Ledieu, P. Unsworth, T.A. Lograsso, A.R. Ross, R. McGrath, Surf. Sci. 600 (2006) 4752.

[78] J. Ledieu, P. Unsworth, T.A. Lograsso, A.R. Ross, R. McGrath, Phys. Rev. B 73 (2006) 12204; J.-N. Longchamp, M. Erbudak, Y. Weisskopf, J. Phys. IV France 132 (2006) 117.

[79] M. Shimoda, J.Q. Guo, T.J. Sato, A.-P. Tsai, J. Non-Cryst. Solids 334 & 335 (2004) 505.

[80] H.R. Sharma, M. Shimoda, A.R. Ross, T.A. Lograsso, A.P. Tsai, Phys. Rev. B 72 (2005) 45428.

[81] N. Ferralis, R.D. Diehl, K. Pussi, M. Lindroos, I. Fisher, C.J. Jenks, Phys. Rev. B 69 (2004) 75410.

[82] K. Saito, K. Ichioka, S. Sugawara, Phil. Mag. 85 (2005) 3629.

[83] E.J. Widjaja, L.D. Marks, Phil. Mag. Lett. 83 (2003) 47.

[84] V.E. Dmitrienko, S.B. Astaf'ev, Phys. Rev. Lett. 75 (1995) 1538.

[85] W. Steurer, Mater. Sci. Eng. 294 − 296 (2000) 268.

[86] E.J. Widjaja, L.D. Marks, Phys. Rev. B 68 (2003) 134211.

[87] J.V. Barth, G. Costantini, K. Kern, Nature 437 (2005) 671.

[88] T. Ito, S. Okazaki, Nature 406 (2000) 1027.

[89] D.M. Eigler, E.K. Schweizer, Nature 344 (1990) 524; L. Bartels, G. Meyer, K.-H. Rieder, Phys. Rev. Lett. 79 (1997) 697.

[90] M. Hehn, K. Ounadjela, J.-P. Bucher, F. Rousseaux, D. Decanini, B. Bartenlian, C. Chappert, Science 272 (1996) 1782; P. Gambardella, A. Dallmeyer, K. Maiti, M.C. Malagoli, W. Eberhardt, K. Kern, C. Carbone, Nature 416 (2002) 301.

[91] R. Allenspach, J. Magn. Magn. Mater. 129 (1994) 160.

[92] C. Stamm, F. Marty, A. Vaterlaus, V. Weich, S. Egger, U. Maier,
      U. Ramsperger, H. Fuhrmann, D. Pescia, Science 282 (1998) 449.

[93] F.J. Himpsel, J.E. Ortega, G.J. Mankey, R.F. Willis, Adv. Phys. 47 (1998)
      511.

[94] S. Blundell, Magnetism in condensed matter, Oxford University Press, Ox-
      ford, 2001, p. 170.

[95] Model HPT-RX 2515, Fisons Instruments Vacuum Generators, Hastings,
      TN38 9NN, GB.

[96] Button Heater, HeatWave Labs. Inc., Watsonville, CA 95076, USA.

[97] A. Zangwill, Physics at surfaces, Cambridge University Press, Cambridge,
      1988.

[98] G. Ertl, J. Küppers, Low energy electrons and surface chemistry, VCH,
      Weinheim, 1985.

[99] S. Tanuma, C.J. Powell, D.R. Penn, Surf. Interface Anal. 17 (1991) 911.

[100] M.P. Seah, W.A. Dench, Surf. Interface Anal. 1 (1979) 2.

[101] H. Li, B.P. Tonner, Surf. Sci. 237 (1990) 141; M. Hochstrasser, M. Zurkirch,
      E. Wetli, D. Pescia, M. Erbudak, Phys. Rev. B 50 (1994) 17705.

[102] M. Henzler, Surf. Rev. Lett. 4 (1997) 489.

[103] M. Erbudak, M. Hochstrasser, T. Schulthess, E. Wetli, Phil. Mag. Lett. 68
      (1993) 179; M. Erbudak, M. Hochstrasser, E. Wetli, M. Zurkirch, Surf. Rev.
      Lett. 4 (1997) 179.

[104] M. Erbudak, T. Schulthess, E. Wetli, Phys. Rev. B 49 (1994) 6316.

[105] S. Kikuchi, Jpn. J. Phys. 5 (1928) 83.

[106] Model BDL800IR, OCI Vacuum Microengineering, London, Ontario N5W
      4R3, Canada.

[107] Model ST-7, Santa Barbara Instruments Group, Santa Barbara, CA 93198,
      USA.

[108] M.J. Freiser, IEEE Trans. Magn. 4 (1968) 152; Z.Q. Qiu, S.D. Bader, Rev.
      Sci. Instrum. 71 (2000) 1243.

[109] Model 110 SFA, Aerotech, Nürnberg, D 90449, Germany.

[110] Model PEM-90, Hinds Instruments, Hillsboro, OR 97124, USA.

[111] H.C. Siegmann, J. Phys.: Condens. Matter 4 (1992) 8395.

[112] Z. Shen, W. Raberg, M. Heinzig, C.J. Jenks, V. Fournée, M.A. van Hove, T.A. Lograsso, D. Delaney, T. Cai, P.C. Canfield, I.R. Fisher, A.I. Goldman, M.J. Kramer, P.A. Thiel, Surf. Sci. 450 (2000) 1.

[113] A. Guinier, X-ray diffraction in crystals — In crystals, imperfect crystals, and amorphous bodies, Freeman, San Francisco, 1993, p. 121.

[114] A.J. Bradley, G.C. Seager, J. Inst. Metals 64 (1939) 81.

[115] P. Villars, L.D. Calvert, Pearson's handbook of crystallographic data for intermetallic phases, ASM, Ohio, 1991.

[116] M. Gierer, M.A. van Hove, A.I. Goldman, Z. Shen, S.-L. Chang, C.J. Jenks, C.-M. Zhang, P.A. Thiel, Phys. Rev. Lett. 78 (1997) 467; M. Gierer, M.A. van Hove, A.I. Goldman, Z. Shen, S.-L. Chang, P.J. Pinhero, C.J. Jenks, J.W. Anderegg, C.-M. Zhang, P.A. Thiel, Phys. Rev. B 57 (1998) 7628.

[117] R. Bastasz, J. Alloys Comp. 342 (2002) 427.

[118] N.R. Shivaparan, M.A. Teter, R.J. Smith, Surf. Sci. 476 (2001) 152.

[119] G.A. Prinz, Phys. Rev. Lett. 54 (1985) 1051.

[120] C.P. Wang, S.C. Wu, F. Jona, P.M. Marcus, Phys. Rev. B 49 (1994) 17385.

[121] H. Wieldraaijer, J.T. Kohlhepp, P. LeClair, K. Ha, W.J.M. de Jonge, Phys. Rev. B 67 (2003) 224430.

[122] D.K. Yu, H.P. Bonzel, M. Scheffler, New J. Phys. 8 (2006) 65.

[123] L. Dong, D.J. Srolovitz, Appl. Phys. Lett. 75 (1999) 584.

[124] G.N. van Wyk, H.J. Smith, Nucl. Instrum. Methods 170 (1980) 433; D. Dobrev, Thin Solid Films 92 (1982) 41.

[125] T. Flückiger, M. Erbudak, A. Hensch, Y. Weisskopf, M. Hong, A.R. Kortan, Surf. Interface Anal. 34 (2002) 441.

[126] A.R. Smith, K.-J. Chao, Q. Niu, C.-K. Shih, Science 273 (1996) 226.

[127] P. Moras, W. Theis, L. Ferrari, S. Gardonio, J. Fujii, K. Horn, C. Carbone, Phys. Rev. Lett. 96 (2006) 156401.

[128] S.D. Bader, J. Magn. Magn. Mater. 100 (1991) 440.

[129] J. Donohue, The structure of the elements, Wiley, New York, 1974.

[130] Z.Q. Wang, Y.S. Li, F. Jona, P.M. Marcus, Solid State Commun. 61 (1987) 623.

[131] C.S. Tian, D. Qian, D. Wu, R.H. He, Y.Z. Wu, W.X. Tang, L.F. Yin, Y.S. Shi, G.S. Dong, X.F. Jin, X.M. Jiang, F.Q. Liu, H.J. Qian, K. Sun, L.M. Wang, G. Rossi, Z.Q. Qiu, J. Shi, Phys. Rev. Lett. 94 (2005) 137210.

[132] Y. Tan, T. Shinoda, Y. Mishima, T. Suzuki, Mater. Trans. 42 (2001) 464.

[133] M. Hansen, Constitution of binary alloys, McGraw-Hill, New York, 1958.

[134] M. Hochstrasser, A. Atrei, B. Bolliger, R. Eismann, M. Erbudak, D. Pescia, Surf. Rev. Lett. 5 (1998) 1007.

[135] R.E. Watson, M. Weinert, Phys. Rev. B 58 (1998) 5981; R. Hultgren, P.D. Desai, D.T. Hawkins, M. Gleiser, K.K. Kelley, Selected values of the thermodynamic properties of binary alloys, ASM, Ohio, 1973.

[136] J.A. Barrow, V. Fournée, A.R. Ross, P.A. Thiel, M. Shimoda, A.P. Tsai, Surf. Sci. 539 (2003) 54.

[137] A. Yamamoto, H. Takakura, A.P. Tsai, Ferroelectrics 305 (2004) 279; H.R. Sharma, M. Shimoda, A.P. Tsai, Jpn. J. Appl. Phys. 45 (2006) 2208.

Acronyms

2f	twofold
3f	threefold
5f	fivefold
10f	tenfold
AES	Auger electron spectroscopy
bcc	body-centred cubic
d-	decagonal
i-	icosahedral
LEED	low-energy electron diffraction
ML	monolayer(s)
MLE	monolayer equivalent(s)
MOKE	magneto-optical Kerr effect
SEI	secondary-electron imaging
SEMPA	scanning electron microscopy with polarisation analysis

Publications

Structure of Gd_2O_3 films epitaxially grown on GaAs(100) and GaN(0001) surfaces
T. Flückiger, M. Erbudak, A. Hensch, Y. Weisskopf, M. Hong, A.R. Kortan
Surf. Interface Anal. 34 (2002) 441

Nanoepitaxy: size selection in self-assembled and oriented Al nanocrystals grown on a quasicrystal surface
T. Flückiger, Y. Weisskopf, M. Erbudak, R. Lüscher, A.R. Kortan
Nano Lett. 3 (2003) 1717

Absorption kinetics of thin Al films on quasicrystalline Al-Pd-Mn
R. Lüscher, M. Erbudak, T. Flückiger, Y. Weisskopf
Appl. Surf. Sci. 233 (2004) 129

Al nanostructures on quasicrystalline Al-Pd-Mn
R. Lüscher, M. Erbudak, Y. Weisskopf
Surf. Sci. 569 (2004) 163

Structural modifications upon deposition of Fe on the icosahedral quasicrystal Al-Pd-Mn
Y. Weisskopf, R. Lüscher, M. Erbudak
Surf. Sci. 578 (2005) 35

Crystalline textures on the Al-Ni-Co quasicrystal surface
M. Erbudak, J.-N. Longchamp, Y. Weisskopf
Turk. J. Phys. 29 (2005) 277

In situ formation of a new Al-Pd-Mn-Si quasicrystalline phase on the pentagonal surface of the Al-Pd-Mn quasicrystal
J.-N. Longchamp, M. Erbudak, Y. Weisskopf
J. Phys. IV France 132 (2006) 117

Ni deposition on the pentagonal surface of an icosahedral Al-Pd-Mn quasicrystal
Y. Weisskopf, M. Erbudak, J.-N. Longchamp, T. Michlmayr
Surf. Sci. 600 (2006) 2594

Self-assembled nano-structures on the icosahedral Al-Pd-Mn quasicrystal
M. Erbudak, J.-N. Longchamp, Y. Weisskopf
Turk. J. Phys. 30 (2006) 213

Quantum size effects arising from incompatible point-group symmetries
P. Moras, Y. Weisskopf, J.-N. Longchamp, M. Erbudak, P.H. Zhou, L. Ferrari,
C. Carbone
Phys. Rev. B 74 (2006) 121405

The quasicrystal-crystal interface between icosahedral Al-Pd-Mn and deposited Co:
evidence for the affinity of the quasicrystal structure to the CsCl structure
Y. Weisskopf, S. Burkardt, M. Erbudak, J.-N. Longchamp
Surf. Sci., in press

Curriculum Vitae

Last and first name	Weisskopf Yves
Date of birth	February 2, 1978
Citizen of	Pratteln, BL, Switzerland

Education

1983 − 1985	Kindergarten Pratteln
1985 − 1990	Primarschule Pratteln
1990 − 1994	Progymnasium Pratteln, Typus C
1994 − 1997	Gymnasium Muttenz, Maturität Typus C
1998 − 2003	Studies of physics at the Swiss Federal Institute of Technology Zurich
10/2002 − 2/2003	Diploma thesis on *Aluminium deposition on the decagonal quasicrystal* $Al_{70}Co_{15}Ni_{15}$ in the group of Prof. Dr. M. Erbudak, Institute for Solid State Physics, Department of Physics, Swiss Federal Institute of Technology Zurich
10/2003 − 9/2006	PhD student in the research area of *Quasicrystal-crystal interfaces* in the group of Prof. Dr. M. Erbudak, Institute for Solid State Physics, Department of Physics, Swiss Federal Institute of Technology Zurich